ADVANCING REGULATORY SCIENCE FOR MEDICAL COUNTERMEASURE DEVELOPMENT

Workshop Summary

I0896197

Theresa Wizemann, Bruce M. Altevogt, and
Anne B. Claiborne, *Rapporteurs*

Forum on Drug Discovery, Development, and Translation

Forum on Medical and Public Health Preparedness
for Catastrophic Events

Board on Health Sciences Policy

INSTITUTE OF MEDICINE
OF THE NATIONAL ACADEMIES

THE NATIONAL ACADEMIES PRESS
Washington, D.C.
www.nap.edu

THE NATIONAL ACADEMIES PRESS 500 Fifth Street, N.W. Washington, DC 20001

NOTICE: The project that is the subject of this report was approved by the Governing Board of the National Research Council, whose members are drawn from the councils of the National Academy of Sciences, the National Academy of Engineering, and the Institute of Medicine.

This project was supported by contracts between the National Academy of Sciences and Department of Health and Human Services (Contract Nos. N01-OD-4-2139 TO #158, HHSF223001003T, HHSF22301022T TO #24, N01-OD-4-2139 TO #198, HHSP233200900680P, HH5P23320042509X1, and HHSP233200800498P), the American Diabetes Association, the American Society for Microbiology, Amgen Inc., the Association of American Medical Colleges, Bristol-Myers Squibb, the Burroughs Wellcome Fund, Celtic Therapeutics, LLLP, the Critical Path Institute, the Doris Duke Charitable Foundation, Eli Lilly & Co., FasterCures, the Foundation for the NIH, Friends of Cancer Research, GlaxoSmithKline, Howard Hughes Medical Institute, Johnson & Johnson, Merck & Co., Inc., Novartis Pharmaceuticals Corporation, Pfizer, Inc, the American College of Emergency Physicians, the American Hospital Association, the American Medical Association, the American Nurses Association, the Association of State and Territorial Health Officials, the Centers for Disease Control and Prevention (Contract No. 200-2005-13434 TO #6), the Department of the Army (Contract No. W81XWH-08-P-0934), the Department of Homeland Security's Office of Health Affairs (Contract No. HSHQDC-07-C-00097), the Department of Homeland Security, Federal Emergency Management Agency (Contract No. HSFEHQ-08-P-1800), the Department of Transportation's National Highway Traffic Safety Administration (DTNH22-10-H-00287) the Department of Veterans Affairs (Contract No. V101(93)P-2136 TO #10), the Emergency Nurses Association, the National Association of Chain Drug Stores, the National Association of County and City Health Officials, the National Association of Emergency Medical Technicians, the Pharmaceutical Research and Manufacturers of America, The Robert Wood Johnson Foundation, and the United Health Foundation. Any opinions, findings, conclusions, or recommendations expressed in this publication are those of the authors and do not necessarily reflect the view of the organizations or agencies that provided support for this project.

International Standard Book Number-13: 978-0-309-21490-2
International Standard Book Number-10: 0-309-21490-4

Additional copies of this report are available from the National Academies Press, 500 Fifth Street, N.W., Lockbox 285, Washington, DC 20055; (800) 624-6242 or (202) 334-3313 (in the Washington metropolitan area); Internet, http://www.nap.edu.

For more information about the Institute of Medicine, visit the IOM home page at: **www.iom.edu.**

Printed in the United States of America

The serpent has been a symbol of long life, healing, and knowledge among almost all cultures and religions since the beginning of recorded history. The serpent adopted as a logotype by the Institute of Medicine is a relief carving from ancient Greece, now held by the Staatliche Museen in Berlin.

Suggested citation: IOM (Institute of Medicine). 2011. *Advancing Regulatory Science for Medical Countermeasure Development: Workshop Summary.* Washington, DC: The National Academies Press.

*"Knowing is not enough; we must apply.
Willing is not enough; we must do."*
—Goethe

INSTITUTE OF MEDICINE
OF THE NATIONAL ACADEMIES

Advising the Nation. Improving Health.

THE NATIONAL ACADEMIES

Advisers to the Nation on Science, Engineering, and Medicine

The **National Academy of Sciences** is a private, nonprofit, self-perpetuating society of distinguished scholars engaged in scientific and engineering research, dedicated to the furtherance of science and technology and to their use for the general welfare. Upon the authority of the charter granted to it by the Congress in 1863, the Academy has a mandate that requires it to advise the federal government on scientific and technical matters. Dr. Ralph J. Cicerone is president of the National Academy of Sciences.

The **National Academy of Engineering** was established in 1964, under the charter of the National Academy of Sciences, as a parallel organization of outstanding engineers. It is autonomous in its administration and in the selection of its members, sharing with the National Academy of Sciences the responsibility for advising the federal government. The National Academy of Engineering also sponsors engineering programs aimed at meeting national needs, encourages education and research, and recognizes the superior achievements of engineers. Dr. Charles M. Vest is president of the National Academy of Engineering.

The **Institute of Medicine** was established in 1970 by the National Academy of Sciences to secure the services of eminent members of appropriate professions in the examination of policy matters pertaining to the health of the public. The Institute acts under the responsibility given to the National Academy of Sciences by its congressional charter to be an adviser to the federal government and, upon its own initiative, to identify issues of medical care, research, and education. Dr. Harvey V. Fineberg is president of the Institute of Medicine.

The **National Research Council** was organized by the National Academy of Sciences in 1916 to associate the broad community of science and technology with the Academy's purposes of furthering knowledge and advising the federal government. Functioning in accordance with general policies determined by the Academy, the Council has become the principal operating agency of both the National Academy of Sciences and the National Academy of Engineering in providing services to the government, the public, and the scientific and engineering communities. The Council is administered jointly by both Academies and the Institute of Medicine. Dr. Ralph J. Cicerone and Dr. Charles M. Vest are chair and vice chair, respectively, of the National Research Council.

www.national-academies.org

PLANNING COMMITTEE FOR THE WORKSHOP ON ADVANCING REGULATORY SCIENCE FOR MEDICAL COUNTERMEASURE DEVELOPMENT[1]

LESLIE Z. BENET *(Co-Chair),* University of California-San Francisco
JOHN REX, *(Co-Chair),* AstraZeneca, Waltham, MA
MARGARET ANDERSON, FasterCures, Washington, DC
PHYLLIS ARTHUR, Biotechnology Industry Organization, Washington, DC
ANTHONY BAHINSKI, Wyss Institute for Biologically Inspired Engineering at Harvard University, Boston, MA
LUCIANA BORIO, Food and Drug Administration, Silver Spring, MD
GAIL H. CASSELL, Harvard Medical School (visiting), Carmel, IN
GARRET A. FITZGERALD, University of Pennsylvania School of Medicine, Philadelphia
STEVEN K. GALSON, Amgen Inc., Thousand Oaks, CA
JOHN GRABENSTEIN, Merck & Co., Inc., West Point, PA
ROBERT KADLEC, PRTM Management Consultants, Washington, DC
MICHAEL KATZE, University of Washington, Seattle, WA
GEORGE KORCH, Department of Health and Human Services, Washington, DC
MICHAEL KURILLA, National Institute of Allergy and Infectious Diseases, Bethesda, MD
BERNARD MUNOS, InnoThink, Indianapolis, IN
GERALD PARKER, Department of Defense, Washington, DC
MARY PENDERGAST, Pendergast Consulting, Washington, DC
NICKI PESIK, Centers for Disease Control and Prevention, Atlanta, GA
ERIC TONER, Center for Biosecurity of the UPMC, Baltimore, MD

IOM Staff

ANNE B. CLAIBORNE, Director, Forum on Drug Discovery, Development, and Translation
BRUCE M. ALTEVOGT, Director, Forum on Medical and Public Health Preparedness for Catastrophic Events
ELIZABETH F. C. TYSON, Research Associate
ANDREW M. POPE, Director, Board on Health Sciences Policy
DONNA L. RANDALL, Administrative Assistant
ROBIN GUYSE, Senior Program Assistant
ALEX REPACE, Senior Program Assistant

[1] Institute of Medicine planning committees are solely responsible for organizing the workshop, identifying topics, and choosing speakers. The responsibility for the published workshop summary rests with the workshop rapporteurs and the institution.

FORUM ON DRUG DISCOVERY, DEVELOPMENT, AND TRANSLATION[1]

[1] Institute of Medicine forums and roundtables do not issue, review, or approve individual documents. The responsibility for the published workshop summary rests with the workshop rapporteurs and the institution.

ELIZABETH (BETSY) MYERS, Doris Duke Charitable Foundation, New York, NY
JOHN ORLOFF, Novartis Pharmaceuticals Corporation, East Hanover, NJ
AMY PATTERSON, National Institutes of Health, Bethesda, MD
MICHAEL ROSENBLATT, Merck & Co., Inc., Whitehouse Station, NJ
JANET SHOEMAKER, American Society for Microbiology, Washington, DC
ELLEN SIGAL, Friends of Cancer Research, Arlington, VA
ELLEN STRAHLMAN, GlaxoSmithKline, Research Triangle Park, NC
NANCY SUNG, Burroughs Wellcome Fund, Research Triangle Park, NC
JORGE A. TAVEL, National Institute of Allergy and Infectious Diseases, Bethesda, MD
JANET TOBIAS, Mt. Sinai School of Medicine, New York, NY
JOANNE WALDSTREICHER, Johnson & Johnson, Raritan, NJ
JANET WOODCOCK, Food and Drug Administration, White Oak, MD
RAYMOND L. WOOSLEY, The Critical Path Institute, Tucson, AZ

IOM Staff

ANNE B. CLAIBORNE, Forum Director
REBECCA A. ENGLISH, Associate Program Officer
ELIZABETH F. C. TYSON, Research Associate
ANDREW M. POPE, Director, Board on Health Sciences Policy
ROBIN GUYSE, Senior Program Assistant

FORUM ON MEDICAL AND PUBLIC HEALTH PREPAREDNESS FOR CATASTROPHIC EVENTS[1]

ROBERT KADLEC *(Co-Chair)*, PRTM Management Consultants, Washington, DC
LYNNE KIDDER *(Co-Chair)*, Center for Excellence in Disaster Management and Humanitarian Assistance, Tripler AMC, Hawaii
ALEX ADAMS, National Association of Chain Drug Stores Foundation, Alexandria, VA
DAMON ARNOLD, Association of State and Territorial Health Officials, Arlington, VA
GEORGES BENJAMIN, American Public Health Association, Washington, DC
VICTORIA DAVEY, Department of Veterans Affairs, Washington, DC
JEFFREY DUCHIN, Seattle & King County and University of Washington, Seattle
ALEXANDER GARZA, Department of Homeland Security, Washington, DC
JULIE GERBERDING, Merck Vaccines, West Point, PA
LEWIS GOLDFRANK, New York University Medical Center, New York
DAN HANFLING, Inova Health System, Falls Church, VA
JACK HERRMANN, National Association of County and City Health Officials, Washington, DC
JAMES JAMES, American Medical Association, Chicago, IL
JERRY JOHNSTON, National Association of Emergency Medical Technicians, Mt. Pleasant, IA
BRIAN KAMOIE, The White House, Washington, DC
LISA KAPLOWITZ, HHS Office of the Assistant Secretary for Preparedness and Response, Washington, DC
ALI KHAN, Centers for Disease Control and Prevention, Atlanta, GA
MICHAEL KURILLA, National Institute of Allergy and Infectious Diseases, Bethesda, MD
JAYNE LUX, National Business Group on Health, Washington, DC
ANTHONY MACINTYRE, American College of Emergency Physicians, Washington, DC
ANGELA McGOWAN, Robert Wood Johnson Foundation, Princeton, NJ
MARGARET McMAHON, Emergency Nurses Association, Williamstown, NJ
MATTHEW MINSON, Texas A&M University, College Station, TX

[1] Institute of Medicine forums and roundtables do not issue, review, or approve individual documents. The responsibility for the published workshop summary rests with the workshop rapporteurs and the institution.

ERIN MULLEN, Pharmaceutical Research and Manufacturers of America, Washington, DC
CHERYL PETERSON, American Nurses Association, Silver Spring, MD
STEVEN PHILLIPS, National Library of Medicine, Bethesda, MD
ROSLYNE SCHULMAN, American Hospital Association, Washington, DC
RICHARD SERINO, Federal Emergency Management Agency, Washington, DC
MICHAEL SKIDMORE, Office of Assistant Secretary of Defense for Health Affairs, Department of Defense, Washington, DC
SHARON STANLEY, American Red Cross, Washington, DC
ERIC TONER, University of Pittsburgh Medical Center, Pittsburgh, PA
REED TUCKSON, UnitedHealth Group, Minneapolis, MN
MARGARET VANAMRINGE, The Joint Commission, Washington, DC
GAMUNU WIEJETUNGE, National Highway Traffic Safety Administration, Washington, DC

IOM Staff

BRUCE M. ALTEVOGT, Forum Director
CLARE STROUD, Program Officer
KRISTIN VISWANATHAN, Research Associate
ANDREW M. POPE, Director, Board on Health Sciences Policy
ALEX REPACE, Senior Program Assistant

Reviewers

This report has been reviewed in draft form by individuals chosen for their diverse perspectives and technical expertise, in accordance with procedures approved by the National Research Council's Report Review Committee. The purpose of this independent review is to provide candid and critical comments that will assist the institution in making its published report as sound as possible and to ensure that the report meets institutional standards for objectivity, evidence, and responsiveness to the study charge. The review comments and draft manuscript remain confidential to protect the integrity of the process. We wish to thank the following individuals for their review of this report:

Gigi Kwik Gronvall, Center for Biosecurity of UPMC
Richard Jaffe, ANSER/Analytic Services Inc.
Jon Mogford, Defense Sciences Office, DARPA
Raymond L. Woosley, The Critical Path Institute

Although the reviewers listed above have provided many constructive comments and suggestions, they were not asked to endorse the final draft of the report before its release. The review of this report was overseen by **Peter Honig,** AstraZeneca. Appointed by the Institute of Medicine, he was responsible for making certain that an independent examination of this report was carried out in accordance with institutional procedures and that all review comments were carefully considered. Responsibility for the final content of this report rests entirely with the authors and the institution.

Contents

Tables, Figures, and Boxes

TABLES

FIGURES

BOXES

Acronyms

ASPR	HHS Assistant Secretary for Preparedness and Response
BARDA	Biomedical Advanced Research and Development Authority
BUN	blood urea nitrogen
CBER	FDA Center for Biologics Evaluation and Research
CBRN	chemical, biological, radiological, nuclear
CDC	Centers for Disease Control and Prevention
CDER	FDA Center for Drug Evaluation and Research
CDISC	Clinical Data Interchange Standards Consortium
CDRH	FDA Center for Devices and Radiological Health
CNS	central nervous system
C-Path	Critical Path Institute
DARPA	Defense Advanced Research Projects Agency
DoD	Department of Defense
EUA	emergency use authorization
FDA	U.S. Food and Drug Administration
GLP	good laboratory practice
GMP	good manufacturing practice

HHS	U.S. Department of Health and Human Services
HL7	Health Level Seven International
IND	investigational new drug
IOM	Institute of Medicine
IRB	institutional review board
LRN	CDC Laboratory Response Network
MBF	mobile bioprocessing facility
MCM	medical countermeasure
MIMIC	Modular IMmune In Vitro Constructs
MRSA	methicillin-resistant *Staphylococcus aureus*
NBSB	National Biodefense Science Board
NCTR	FDA National Center for Toxicological Research
NIAID	National Institute of Allergy and Infectious Diseases
NIH	National Institutes of Health
NRC	National Research Council
PCAST	President's Council of Advisors on Science and Technology
PCR	polymerase chain reaction
PHEMCE	Public Health Emergency Medical Countermeasures Enterprise
PK/PD	pharmacokinetic/pharmacodynamic
PSTC	Predictive Safety Testing Consortium
RNA	ribonucleic acid
SNS	Strategic National Stockpile
VAERS	Vaccine Adverse Event Reporting System
VXDS	Voluntary Exploratory Data Submission program

1

Introduction[1]

During public health emergencies such as influenza pandemics or chemical, biological, radiological/nuclear (CBRN) attacks, safe and effective vaccines, drugs, diagnostics, and other medical countermeasures (MCMs) are essential to protecting national security and the well-being of the public. The U.S. Food and Drug Administration (FDA) plays a central and crucial role in domestic preparedness through its regulation of drugs, biologics, medical devices, and radiation-emitting products. However, recent reports have highlighted the need for improved regulatory science to help address the novel regulatory issues the agency faces in its review and approval of many MCMs (e.g., ethical or practicality barriers to conducting standard clinical trials, balancing benefit and risk for products that would be used only under dire circumstances) (FDA, 2007; NBSB, 2010). A report from the National Biodefense Science Board concluded that

> FDA has not been able to fulfill its implicit national security mission, in large part because of a lack of resources. . . . It is imperative for America's health and progress for FDA to be provided adequate resources to bring

[1] This workshop was organized by an independent planning committee whose role was limited to identification of topics and speakers. This workshop summary was prepared by the rapporteurs as a factual summary of the presentations and discussions that took place at the workshop. Statements, recommendations, and opinions expressed are those of individual presenters and participants and are not necessarily endorsed or verified by the Forums or the National Academies, and they should not be construed as reflecting any group consensus.

its regulatory science into the 21st century. Doing so will greatly enhance FDA's ability to support MCM development and licensing. (NBSB, 2010, pp. 43–44)

In August 2010, the U.S. Department of Health and Human Services (HHS) released its review of the Public Health Emergency Medical Countermeasures Enterprise (PHEMCE) and made numerous recommendations to transform the MCM enterprise to increase its speed, agility, capacity, and success rate, including the promotion of regulatory innovation and investment in regulatory science at FDA (ASPR, 2010).[2] In this regard, FDA has established an MCM initiative that includes (1) enhancing the regulatory review process for the highest-priority MCMs and related technologies; (2) advancing regulatory science to support MCM development and evaluation; and (3) modernizing the legal and regulatory framework to support public health preparedness and response.

At the request of FDA, the Institute of Medicine's (IOM's) Forum on Drug Discovery, Development, and Translation and the IOM's Forum on Medical and Public Health Preparedness for Catastrophic Events jointly convened a workshop on March 29–30, 2011. The workshop, titled Advancing Regulatory Science for Medical Countermeasure Development, was designed to (1) examine ways to advance regulatory science for MCM development and regulatory evaluation; (2) identify scientific opportunities to improve, simplify, or speed MCM development; and (3) identify tools and methods to improve the predictability and success rate of candidate MCMs (see Box 1-1).

Workshop speakers and attendees consisted of experts from federal government agencies, the private sector, and academia, who were invited to discuss the applicability of cutting-edge science to MCM discovery and development with the goal of informing the FDA MCM initiative and the regulatory science-based product review process. Speakers were asked to introduce their comments by providing a brief overview of scientific advances or emerging technologies holding promise to facilitate development of MCMs, and then to focus on what regulatory science advances are needed to address gaps in currently available tools to predict and evaluate product safety, efficacy, and quality. They were also asked to identify how innovative regulatory science methodologies can address emerging tech-

[2] As part of the PHEMCE review, and at the request of the Secretary of HHS and the HHS Assistant Secretary for Preparedness and Response, the IOM's Forums on Drug Discovery, Development, and Translation and Medical and Public Health Preparedness for Catastrophic Events collaborated to host a workshop in February 2010, which addressed challenges facing the PHEMCE. Workshop participants discussed federal policies and procedures affecting the research, development, and approval of medical countermeasures and explored opportunities to improve the process and protect Americans' safety and health (IOM, 2010).

BOX 1-1
Workshop Objectives

- Provide a broad overview of current efforts underway at FDA and other agencies within HHS and the Department of Defense (DoD) to support the research, development, evaluation, and production of MCMs.
- Review novel scientific methodologies used by academia and industry to facilitate development of next generation vaccines, biologics, drugs, and devices.
- Identify major gaps in currently available tools to predict and evaluate product safety, efficacy, and quality.
- Identify opportunities for collaboration and coordination with FDA and among relevant federal and industry programs to support the MCM initiative's regulatory science program, and to develop more clearly defined pathways for product approval.
- Identify regulatory science tools and methodologies to address emerging technologies, targets, and novel products as well as innovative approaches for predicting safety and efficacy (e.g., biomarkers, *in silico* modeling).

nologies, targets, and novel products. Each speaker was asked to identify the top 2–3 regulatory science needs or priorities, and to comment on partnerships or collaborations that could serve as models for, or facilitate achieving, this regulatory science agenda.

Invited discussants were asked to provide brief remarks during the context of a panel discussion to offer a case study or example illuminating success stories or lessons learned with respect to the subject of the panel discussion in which they were invited to participate.

This workshop summary identifies key issues and raises awareness of opportunities and challenges in advancing regulatory science underpinning regulatory decision making about MCM products. After first providing context and background, the summary presents enterprise stakeholder perspectives (federal and private sector) on the key regulatory science needs and priorities to advance MCM development (Chapter 2); highlights novel science as well as regulatory science tools to address that novel science (Chapter 3); and discusses challenges in applying regulatory science to MCM development specific to at-risk populations such as children and pregnant women (Chapter 4). Chapters 5 and 6 summarize crosscutting themes and provide concluding remarks.

The summary provides an overview of the highlights from the substantial discussions that took place at the workshop. A key goal of the workshop was to identify regulatory science tools and methods that are available or under development, as well as major gaps in currently available regulatory science tools. This summary report includes key and crosscutting themes and compiles a number of suggestions offered by workshop attendees. Throughout the workshop a number of participants noted that further work, including meetings and formation of working groups, will be important to drill down more deeply in each scientific area to fill out a more robust regulatory science agenda and priorities with respect to each area of science.

DEFINING REGULATORY SCIENCE

FDA defines regulatory science as "the science of developing new tools, standards, and approaches to assess the safety, efficacy, quality and performance of FDA-regulated products" (FDA, 2010). Jesse Goodman, chief scientist and deputy commissioner for science and public health at FDA, emphasized that regulatory science is a critical bridge between basic science discoveries and early work in industry and the approval of new products that can help patients. With regard to MCMs, enhanced regulatory science is needed to reduce uncertainty and provide clear regulatory pathways for MCM development. Exercising the principles of regulatory science requires underlying capacity and expertise at FDA, Goodman added.

George Korch, acting principal deputy assistant secretary for preparedness and response, HHS, expanded on the definition of regulatory science given by Goodman, suggesting that in context of the PHEMCE review, it means specifically that:

- Regulatory pathways for difficult issues in development of medical products against CBRN and emerging infectious disease threats are better defined.
- Industry, the Biomedical Advanced Research and Development Authority (BARDA), the National Institutes of Health (NIH), and the regulatory community work in partnership to identify and prioritize the most urgent needs.
- The commercial sector helps FDA where it can, in gaining access to new technologies that ultimately need regulatory oversight for product approvals.
- All centers at FDA benefit from research investment, and all centers work collaboratively to maximize the impact of these funds in accelerating the pace of product approvals for PHEMCE-related countermeasures.

- The PHEMCE partners[3] actively support FDA's efforts to continue to benefit from investments made in regulatory science.

CHARGE TO THE WORKSHOP PARTICIPANTS

In his keynote address and charge to the workshop participants, Goodman reviewed some of the challenges of MCM development, described the framework for the FDA Regulatory Science Initiative, and listed key questions FDA is seeking expert input on (summarized in Box 1-2).

Goodman also emphasized the importance of engaging the agency early in the MCM development process, so FDA regulatory science can be brought to bear in a highly interactive manner with the product sponsor. He remarked that there is a need for broad collaboration, as no single agency or entity has all the necessary tools or expertise to meet the challenge of bringing MCMs to market. He noted that a component of the FDA MCM initiative is aimed at facilitating engagement between sponsors and FDA early in development, working together to identify where the gaps are, and coming to general agreement on what models, outcomes, and data are needed to allow for emergency use of a product, and he remarked that this model of engagement is different than how commercial drug development is normally done.

IMPLEMENTING THE 2010 PHEMCE REVIEW

As noted above, in the summer of 2010, HHS released its review of the MCM enterprise (ASPR, 2010). Korch quoted HHS Secretary Kathleen Sebelius' vision for the MCM enterprise as follows:

> Our nation must have the nimble, flexible capacity to produce MCMs rapidly in the face of any attack or threat, known or unknown, including a novel, previously unrecognized naturally occurring emerging infectious disease. (ASPR, 2010, p. 6)

In conducting the review of the MCM enterprise, Korch said, several major areas of risk for MCM product developers were identified:

- *technical risk* for product development (e.g., access to core manufacturing and accessory services, limited or no ability for translation and incubation of promising ideas from technology base to early product development);

[3] The PHEMCE partners comprise certain operating divisions within HHS (CDC, FDA, NIH, BARDA) and other partners in the Departments of Defense, Homeland Security, Agriculture, and Veterans Affairs.

- *regulatory risk* (e.g., uncertain or high-risk pathway for product licensure, need for greater resources at FDA to mitigate risk profile);
- *business risk* (e.g., financial pressures for capital for start-up or maturing businesses, lack of integrated business expertise in new organizations); and
- *governmental risks* (e.g., deficient coordination across multiple agency and departmental programs; lack of ability to project and prioritize long-range needs through an integrated multiyear program and budget process).

Based on the identified barriers and challenges, the review highlighted a variety of opportunities for improvement, the first of which

BOX 1-2
Highlights of Keynote Address

CHALLENGES TO MCM DEVELOPMENT

- Increasing costs of product development
- Low success rates
- Uncertainties throughout the process
- Uncertain market for MCM products
- Highly variable time frames for product development, production, and mobilization in response to an emergency
- Limited application, thus far, of systems biology, new technologies, and high-level computational science approaches to product development (approaches that are routinely applied at the basic science level)

FRAMEWORK FOR THE FDA REGULATORY SCIENCE INITIATIVE

- Leadership and coordination across the agency, and development of an agency-wide strategic science and innovation plan
- Support for applied regulatory science research, both within FDA and through collaborations
- Scientific and professional development
- Recruitment and retention of highly qualified personnel in key areas

QUESTIONS TO BE ADDRESSED

- What tools and capacity are needed to develop and evaluate the products of today?
- What will be needed to evaluate products and technologies on the horizon?
- How can uncertainty around platform technologies be reduced, so the risk of failure in an emergency is reduced?
- How can collaboration be improved? How can FDA better take advantage of resources across the government and the private sector?
- How should FDA prioritize its scientific work if presented with additional resources? In the absence of new resources, what areas are most critical?

is enhancing regulatory innovation, science, and capacity at FDA. Recommendations for optimizing the enterprise were made as well. Korch highlighted five key initiatives recommended in the report: major investments in upgrading science at FDA; establishing Centers of Excellence in Advanced Development and Manufacturing; expanding the pipeline at the National Institute of Allergies and Infectious Diseases (NIAID); addressing the immediate needs for influenza vaccines (e.g., sterility, potency, optimization); and establishing a strategic investment fund to increase government investment in entrepreneurial commercial ventures. He noted that this IOM workshop is directly related to the first initiative on upgrading regulatory science at FDA.

In summary, Korch said, the U.S. government has made a tremendous investment over the last 10 years, and there are many successes and capabilities that did not exist a decade ago, but gaps remain. The PHEMCE review addressed gaps in the ability of the government to increase the product pipeline and lower the risks for commercial partners, and focused on the development of a capabilities-based (rather than threat-based) strategy for medical defense. "You get the system that you reward," Korch said, so moving forward, the MCM enterprise needs to reward flexibility, fresh thinking, and a vision of how to prepare against low-probability, high-consequence events.

OVERVIEW OF THE FDA MCM INITIATIVE

The mission of the FDA MCM initiative is to promote the development and availability of MCMs by establishing clear regulatory pathways based on the best available science, explained Luciana Borio, acting director of the Office of Counterterrorism and Emerging Threats at FDA. To achieve this, FDA is aggressively taking action in three major areas, referred to as the "pillars" of the initiative:

- Pillar 1: Enhance the MCM review process;
- Pillar 2: Advance regulatory science for MCM development and evaluation; and
- Pillar 3: Optimize the legal, regulatory, and policy framework to support preparedness and response.

In addition to increasing review capacity and expertise in the medical products centers, Pillar 1 will establish public health and security action teams to support FDA reviewers. Teams may, for example, provide targeted briefings for reviewers on threats and risks associated with high-priority MCMs. Although established under Pillar 1, the teams are charged with considering the whole range of regulatory, scientific, and policy issues facing MCM development and approval. The first action team launched, for example, is helping to develop novel regulatory approaches to address the complex regulatory challenges associated with multiplex in vitro diagnostic tests for infectious diseases (which may detect hundreds of agents in a single assay).

When the scientific foundations that underpin regulatory assessments are immature, as they often are in the case of MCMs, Borio said, product development suffers. The Pillar 2 regulatory science program seeks to develop solutions to complex scientific regulatory problems, identify situations in which the application of new science could simplify or speed product development and review, and establish FDA capacity to meet high-priority public health needs, especially during emergencies. Pillar 2 regulatory science was the topic of this workshop.

Borio emphasized that to achieve these goals, FDA will need to access all available expertise and leverage both FDA and non-FDA resources. As such, the MCM initiative involves internal collaborative research, as well as partnerships with other U.S. government agencies, academia, and industry.[4] FDA has stood up internal regulatory science research projects to support product development, approval, and use by making available to product review teams the most up-to-date science-based methods available for regulatory assessment of MCMs. Projects are being funded through internal FDA funds under an interactive peer-review process involving enterprise partners, including NIAID, BARDA, and the Department of Defense (DoD), which helps ensure alignment with MCM enterprise priorities. The discussions at this IOM workshop will help to refine FDA internal programs and identify opportunities for collaborations and partnerships, she said.

SUMMARY OF KEY WORKSHOP THEMES

The workshop was structured in three broad modules. First, stakeholders in the MCM enterprise (public and private) presented their views on the current state of MCM regulatory science and the needs and opportunities for development of MCM regulatory science. Second, a series of panel discussions, anchored by presentations, was held that reviewed novel scientific methodologies in MCM development and identified high-priority MCM regulatory science needs and opportunities. Third, in a summary discussion session, themes, priorities, and future directions were identified by the workshop participants. Over the course of the

[4] Borio also noted that use of funding for the MCM initiative is currently restricted to pandemic influenza activities. FDA is working with congressional leaders to remedy this to allow FDA to use existing appropriations to broadly implement the strategies of the MCM initiative.

workshop, numerous individual suggestions were made for priorities and future directions for advancing regulatory science for MCM development, and in this process a number of crosscutting themes arose that resounded across panel topics. These themes and "big ideas" were synthesized into a high-level summary discussion facilitated by workshop co-chair John Rex, Vice President of Clinical Infection of AstraZeneca. Due to their crosscutting nature, resonating themes and "big ideas" are listed below as a means of orienting the subsequent workshop summary, which summarizes the discussions underlying and supporting these themes.

The themes and "big ideas" listed below are not inclusive of the many individual suggestions that were made throughout the workshop and are summarized elsewhere in this workshop summary. They are compiled as part of the factual summary of the workshop, and should not be construed as reflecting consensus or endorsement by the workshop, the Forums, or the National Academies.

Themes

- *Incentives* drive the process of development, and "you get what you reward." Reward of flexibility and innovative thinking will advance the MCM enterprise.
- *Education and training* are essential, and there is a need to promote an interdisciplinary regulatory science workforce.
- *Defining metrics of success* is not straightforward. Workshop participants offered a number of potential success indicators, such as development of assays and biomarkers, definition of regulatory or approval pathways, product approvals, and addition of products to the Strategic National Stockpile (SNS).
- The *benefit-risk calculus* may be different for MCMs to be used in low-probability/high-consequence events than for traditional products.
- *Precompetitive collaborations* exist and should be promoted and strengthened; data sharing within FDA should be enhanced.
- There is a need for better understanding of *animal models* and how to apply them.
- *Early engagement* between product developers and FDA is imperative to support effective application of regulatory science to the product development process.
- *Cross-enterprise collaboration*, including FDA, other HHS divisions, DoD, industry, and academia, is essential. There is a need for real-time, ongoing, unimpeded collaboration among product developers, government managers, and regulatory review and FDA research scientists.

- MCM product development should be viewed as an *opportunity to advance the development of regulatory science more broadly*, and these advances could have implications for and influence on the traditional biopharma product development model.

Big Ideas

- Promote "big science" to make assays for all human proteins. To accomplish this there is a need to overcome challenges in statistical design.
- *Animal models*: Build databases of existing animal models (genome to phenome) including information of what is known about animal and human responses for key diseases.
    - Develop partnerships and collaborations for data sharing to enhance knowledge on animal models.
    - Work toward approval of an MCM product based on safety and efficacy data from an animal model but with information on clinical dosing.
- "App" technology for public education and communication, surveillance, response, and adverse event monitoring holds promise.
- With respect to *diagnostics*, there is a need for development of a universal transport matrix to "move the sample" during an event, and development of a local test that allows remote data analysis ("move the data").
- The *promise of new statistics* such as Bayesian-augmented control design was discussed.
- Participants noted that there is a need for universal data standards for common data elements in naming and defining variables.
- To enhance *vaccine* development, it was suggested to create platforms for antigens and adjuvants, with accompanying licensure pathways. Workshop participants also suggested that collaborative working groups be convened to share data and experiences with respect to such things as safety biomarkers and platforms.

Additional individual suggestions made by workshop participants are listed at the end of the subsections in Chapter 3 of this workshop summary.

Urgency of the Need

From the outset of and throughout the workshop many participants emphasized the urgency of the need for action to enhance regulatory science and facilitate the development of safe and effective MCMs. In his keynote address, Goodman remarked that MCMs are a unique product in that they are needed for very specific threats and they must be available on a much shorter timeline than that which is seen in more traditional pharmaceutical development efforts. Goodman added that FDA had requested the workshop be held on a very short timeline due to the need to provide information to the FDA MCM initiative as early as possible. Gerald Parker, deputy assistant secretary to the secretary of defense for chemical and biological defense at the DoD, emphasized that "perhaps complacency may be our biggest threat." He noted that it may be more useful to acknowledge that the issues are "high probability, high consequence" ones, and that there is a need to appreciate the sense of urgency associated with this threat to create the important social drivers needed to effect change.

Sidebar: Regulatory Framework for Review and Approval of MCMs

Although the focus of the workshop was regulatory science, there was also significant discussion of the closely related topic of the regulatory framework for review and approval of MCMs (which is being addressed in Pillar 3 of the FDA MCM initiative). Major regulatory framework themes are summarized in this sidebar.

There is significant intersection and overlap across the three pillars of the FDA MCM initiative. This workshop stems from Pillar 2 and was therefore limited in scope to discussions of advancing FDA regulatory science for MCM. However, science must also be interpreted, and regulatory and policy decisions must be made. Throughout the workshop, participants also offered comments regarding the review process as it relates to MCMs, and the underlying legal, regulatory, and policy framework. The topics most often mentioned by participants were the need for an alternate paradigm for approval of MCMs for disaster situations, and modifications to requirements under the Animal Rule. There were also concerns about the liability of MCM developers.

Approval with Conditions

There was much discussion about the need to adapt the existing regulatory framework to support MCM development prior to an emergency declaration. Eric Rose, CEO and chair of the Board of Directors of SIGA Technologies, Inc., noted that the existing emergency use authorization (EUA)[a] mechanism works well as an FDA response process, but it is not a preparedness mechanism. A number of participants advocated for a "provisional approval" mechanism for novel MCM. Products would meet the existing criteria for use under an EUA but could be provisionally or conditionally licensed prior to an actual emergency declaration. There would be advance determination of what would trigger issuance of an EUA and what monitoring would be required once the product was in use.

continued

Sidebar Continued

Rose defined such products as those that (1) may be effective in meeting an otherwise unmet clinical or public health need posed by the potential emergency, (2) have strong evidence of safety relative to the risks and consequences of the unmet need, and (3) have a reasonable basis for dosimetry. Such a product would not be used until there was a disaster, at which point FDA would conduct one final review of the current situation and authorize its use. Subsequent full approval of provisionally approved MCMs could occur when substantial evidence of effectiveness, robust evidence of safety relative to risks and consequences, and a robust basis of dosimetry were available. Many participants emphasized that the threshold for what constitutes practical efficacy data should be set with the understanding that these products fill an unmet need in a very-high-consequence scenario.

A participant cautioned that, regardless of the terminology used, if a provisionally or conditionally approved or licensed intervention is still considered to be an investigational product, there are statutory barriers to the DoD administering investigational products to troops, and insurance companies tend not to cover products that do not have full FDA approval. It was suggested that instead, FDA already has the authority to approve products with conditions for their safe use. Under this authority, some argued that FDA could impose conditions such as, for example, that the MCM is only ever used under an EUA, only from the SNS, or only for specified needs.

Ed Nuzum, chief of the Biodefense Vaccines and Other Biological Products Development Section at NIAID, suggested that having a defined provisional approval step—a clear and potentially attainable goal short of full approval—could help to incentivize small companies and their investors to develop MCM.

Goodman of FDA noted that FDA is actively discussing these issues and potential mechanisms.

Note: Some participants referred to such a mechanism of provisional approval prior to an emergency as a "pre-EUA." However, Michael Kurilla, director of the Office of BioDefense Research Affairs and associate director of Biodefense Product Development at NIAID, explained that the term *pre-EUA* is used by FDA to refer to a submission from a sponsor regarding potential future emergency use of a product, for the purposes of allowing FDA to begin to formulate draft EUA documentation, so that these drafts could be finalized more quickly should an emergency be declared. He noted that there has been variability as to when in the development of a product it could be considered eligible for data to be submitted for pre-EUA status, and what the pre-EUA package should look like is vague and ill defined. (The EUA itself is not a product approval pathway. Rather, it authorizes the use of an unapproved medical product, or an unapproved use of an approved medical product, during a declared emergency. This is based on FDA's assessment that the data conclude that the known and potential benefits are likely to exceed the known and potential risks.)

The Animal Rule

In addition to discussions of the regulatory science aspects of the Animal Rule[b] (see Chapter 3), participants expressed concerns about the implementation of the rule by FDA. It was noted that there are different interpretations of the rule across FDA divisions—and sometimes between reviewers within the same division—of how to apply to the Animal Rule. As a result, the rule is often inconsistently applied to sponsors. Elizabeth Leffel, director of nonclinical sciences at PharmAthene, said that a strategic plan for utilizing the Animal Rule is needed, starting with finalizing the current draft guidance.

Rose noted in his remarks that the Animal Rule is the primary obstacle to demonstrating substantial effectiveness of a product. He pointed out that many threat agents are not testable in clinical scenarios (one could not conduct a randomized trial of an Ebola treatment, for example), yet the Animal Rule has never been the basis for approval of a new chemical entity. Rose raised concerns that decision making at FDA on the definition of acceptable animal efficacy data for specific products is riddled with delay. He also suggested that FDA approval of new indications for previously approved drugs (e.g., ciprofloxicin approval for anthrax postexposure prophylaxis) sets a precedent for imputing substantial evidence of effectiveness from animal models.

Manufacturer Liability

The Public Readiness and Emergency Preparedness (PREP) Act provides liability coverage for manufacturers whose MCM product is used under an EUA. There is also liability protection for a company that advances a product under contract with BARDA. It was noted, however, that liability concerns persist. For example, if an MCM were approved and subsequently provided to the public in a different manner (e.g., not under an EUA, not from the SNS), liability protection likely would not be available. A participant suggested a legislative change be made to include nonvaccine MCMs under the National Vaccine Injury Compensation Program.

---

[a] Under Section 564 of the Federal Food, Drug, and Cosmetic Act, as amended by Project BioShield Act of 2004, the FDA commissioner may authorize the use of an unapproved medical product, or an unapproved use of an approved medical product, during a declared emergency involving a heightened risk of attack on the public or U.S. military forces, or a significant potential to affect national security. http://www.fda.gov/RegulatoryInformation/Guidances/ucm125127.htm (accessed June 9, 2011).

[b] The Animal Rule (21 CFR 314.600 for drugs; 21 CFR 601.90 for biological products) establishes a regulatory process for FDA approval of certain new drugs and biologics used to reduce or prevent the toxicity of CBRN substances based on animal data, when adequate and well-controlled efficacy studies in humans cannot be ethically conducted and field trials are not feasible. See *Guidance for Industry, Animal Models—Essential Elements to Address Efficacy Under the Animal Rule* (Draft Guidance) http://www.fda.gov/downloads/Drugs/GuidanceComplianceRegulatoryInformation/Guidances/UCM078923.pdf (accessed June 9, 2011).

2

MCM Enterprise and Stakeholder Perspectives

FDA REGULATORY SCIENCE RESEARCH NEEDS

To provide a framework for subsequent discussions, representatives from FDA's Center for Drug Evaluation and Research (CDER), Center for Biologics Evaluation and Research (CBER), and Center for Devices and Radiological Health (CDRH) provided a broad overview of current efforts underway at the agency to support research, development, and evaluation of MCMs, and discussed current challenges and how regulatory science can assist in advancing MCM development and evaluation.

Center for Drug Evaluation and Research (CDER)

The regulatory review process is an ongoing, iterative process, explained Susan McCune, deputy director of the Office of Translational Science at CDER. It is hoped that gaps and questions identified during a regulatory review can be addressed through application of regulatory science and that regulatory science in turn informs the broader regulatory review process.

Within CDER, the Science Prioritization and Review Committee has identified science and research needs across seven key areas. McCune pointed out that although these identified needs are centerwide and not limited to MCMs, all are relevant to the MCM initiative:

- Improve access to postmarket data sources and explore feasibility of their use in different types of analyses.

- Improve risk assessment and management strategies to reinforce the safe use of drugs.
- Evaluate the effectiveness and impact of different types of regulatory communications to the public and other stakeholders.
- Evaluate the links among product quality attributes, manufacturing processes, and product performance.
- Develop and improve predictive models of safety and efficacy in humans.
- Improve clinical trial design, analysis, and conduct.
- Enhance individualization of patient treatment.

McCune briefly reviewed the drug regulatory review life cycle (Figure 2-1) and offered a number of examples of potential areas for Pillar 2 research across the drug life cycle (Table 2-1). CDER scientists are already working in many of these areas. For each of these areas, she said, there are numerous potential scientific studies that could be done. One of the primary challenges is prioritization of studies as they relate to the MCM initiative.

In closing, McCune said, CDER has a robust regulatory science program with significant expertise to support the research agenda of the MCM initiative, and CDER researchers and reviewers are eager to collaborate on efforts to advance the regulatory science needs of the initiative.

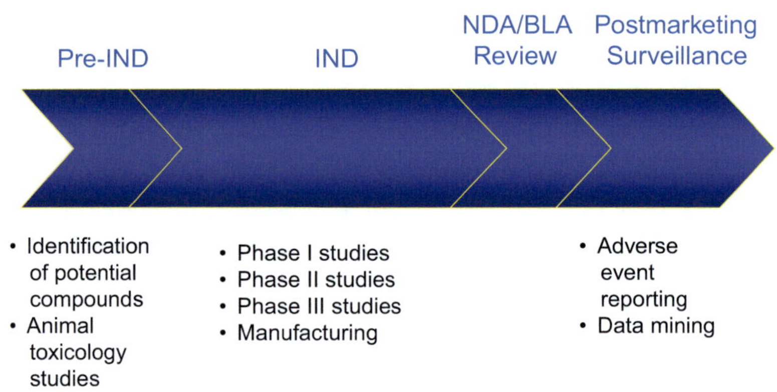

FIGURE 2-1 FDA regulatory review cycle.
NOTE: BLA, biologics license application; IND, investigational new drug application; NDA, new drug application.
SOURCE: Susan McCune. 2011. Presentation at IOM workshop; Advancing Regulatory Science for Medical Countermeasure Development.

Center for Biologics Evaluation and Research (CBER)

Products regulated by CBER include blood, blood components and derivatives, vaccines, allergenic products, cell and gene therapies, xenotransplantation products, human tissues, and various related devices, explained Carolyn Wilson, associate director for research at CBER. MCMs fall into several of these categories. For example, in the area of cell and gene therapies, mesenchymal stem cells are being evaluated for treatment of acute radiation syndrome; pathogen-specific immunoglobulins come under the area of blood components and derivatives; and there are a variety of MCM-related vaccines (e.g., anthrax, botulinum, smallpox, influenza).

Wilson highlighted five areas under Pillar 2 of the MCM initiative in which CBER is conducting research (Table 2-2).[1] The anticipated public health outcomes of such research include the development of new scientific tools and biomarkers to facilitate development of safe and effective MCM biologics and improved guidance to sponsors on how to develop and evaluate MCM biologics, including guidance on how to implement the Animal Rule. Research could result in, for example, earlier identification of toxicity, improved means to assess potential for efficacy, and more rapid detection of safety signals, potentially leading to improved decision making regarding benefits and risk.

One example of current CBER research on animal model biomarkers involves luciferase-expressing *Bacillus anthracis*, which allows for more precise staging of the bacterial infection in mice, and improved design for studies of such things as postexposure prophylaxis and combination therapies. This approach is generalizable, Wilson said, and could be applied to other pathogens and to the development of improved animal models. Other examples of ongoing CBER research mentioned by Wilson include using phage display libraries to evaluate the human immune response, developing faster methods to generate reference reagents for influenza vaccines, and improving detection of adventitious infectious agents in complex samples.

Center for Devices and Radiological Health (CDRH)

The devices regulated by CDRH include diagnostic and detection devices, personal protective equipment (e.g., N95 respirators), emergency devices (e.g., ventilators, intravenous administration sets, resuscitation equipment, drug or vaccine delivery systems, needles), and combination

[1] Further information on CBER biologics research projects can be found at http://www.fda.gov/BiologicsBloodVaccines/ScienceResearch/default.htm (accessed June 9, 2011).

TABLE 2-1 Examples of Potential Areas for Pillar 2 Research in CDER

Phase of Development	Potential CDER Research Areas
Identification of Potential Molecules for Study	Repurposing drugs Molecular modeling Screen approved drugs for other pathogens Effects of combinations of antimicrobials
Animal Toxicology Studies and Animal Models for MCM	Animal models: For pregnancy Modeling of disease states in general To study toxin effects on organ systems Qualified through the drug development tools qualification process For combination added benefit studies For placebo studies For postexposure prophylaxis To evaluate potential safety signals For studies of natural history and pathophysiology of disease Conversion of data from animal model studies to standard format Animal model database
Human Safety Studies for MCMs or Clinical Safety and Efficacy Studies for Influenza	Evaluation of the effects of genetic variations Studies of dosage forms for special populations Extrapolation models from animal to human, including dose scaling for special populations Pediatric safety studies, including ethical issues Understanding human disease through the world literature Development of clinical endpoints for seriously ill influenza patients Development of threat-based data standards Development of standardized case report forms for data collection Modeling drug interaction studies Modeling of PK/PD to labeled drugs for special populations
Manufacturing and Product Quality	Therapeutic protein PK/PD comparability studies Shelf-life extension studies Develop stable product formulations Rapid detection of problems with marketed products

TABLE 2-1 Continued

Phase of Development	Potential CDER Research Areas
Incident-Related Studies	Develop hospital networks for rapid information transfer Communication studies on emergency notification Data mining for adverse events associated with therapy Develop protocols for use during events Real-world use studies on home preparation instructions Real-time epidemiology cluster monitoring

NOTE: PK/PD, pharmacokinetic/pharmacodynamic.

TABLE 2-2 Examples of Potential Areas for Pillar 2 Research in CBER

Pillar 2 Program	Research Needs
Animal Model Biomarker Program	Identify in vitro or in vivo correlates of bioactivity, safety, toxicity Methods development Develop and evaluate animal models
Clinical Biomarker and Immunology Program	How to bridge from animal studies to the human immune response to vaccines Insufficient knowledge of human disease Clinical trial design
Ensuring MCM Product Quality	Measurable product characteristics that correlate with safety and efficacy Improved methods to assess new cell substrates New methods that incorporate new technology and are faster, use fewer animals, and have improved sensitivity/specificity
Radiation Injury Protection and Response Program	Improved tools to assess critical product attributes of cell therapy-based products used to treat radiation injury Animal models to evaluate product safety and potential for efficacy
Health Informatics/ Scientific Computing	Improved means to detect rare adverse events Improved methods and access to health care data sources to monitor safety of marketed MCM biologics Improved use of high-performance computing Tools/models for risk assessment

products (a combination of drug or biological and a device), explained Murray Malin, acting director of the Medical Countermeasure Initiative at CDRH. There are MCM devices in each of these categories. The CDRH Office of In Vitro Diagnostics has developed multiple guidance documents that affect MCMs and has improved transparency by posting device clearance reviews on their website.

Malin highlighted the regulatory science priorities in CDRH, noting that they are similar to those of the other centers:

- develop infrastructure to support development of diagnostics and other MCMs;
- characterize the medical device supply chain;
- enable real-time or near-real-time surveillance of supply, utilization, and availability of medical devices to avoid shortages;
- address special needs populations, point of care, and personalized use of MCMs;
- enhance ability to capture, monitor, and analyze large datasets;
- create statistical tools to develop innovative clinical trials, perform comparative effectiveness research, and perform active surveillance of adverse event reports;
- genomic sequencing devices and assays for the detection of pathogens and antimicrobial drug resistance;
- multiplex/microarray diagnostic devices capable of simultaneous detection/identification of multiple organisms;
- develop new tools to evaluate nanotechnology-based devices, including use as diagnostic markers;
- increase scientific capacity and expertise to prepare for and facilitate new technologies;
- develop guidance for development of multiuse products and platforms to expand MCM product pipelines during emergencies;
- infrastructure, comparator sequencing database, data processing, and resources to enhance reviews of MCMs, facilitate innovative statistical techniques and clinical trials, and develop regulatory pathways for MCMs;
- agreement with Centers for Medicare and Medicaid Services (CMS) on data necessary for Clinical Laboratory Improvement Amendments (CLIA) waiver of authorized products during an emergency; and
- develop high-performance/scientific computing to facilitate the following:
    - answering questions related to comparative effectiveness associated with patient subsets;
    - support for genomic sequencing, multiplex devices;

- identification of improved methods for characterizing failure analysis, validation of factors affecting manufacture of MCMs, and development of forensic evaluation techniques necessary to support multiplatform/product development; and
- development of innovative statistical methods and clinical trial design.

Discussion

Much of the discussion with the FDA panelists focused on communication and collaboration as key components of advancing regulatory science. Wilson pointed out that CBER has "research reviewers" who are actively engaged in both the regulatory science agenda and review activities. Research reviewers can help identify gaps in the science, as well as methods, tools, and reference materials that could help move a technology or a whole product area forward. McCune added that many CDER review team members are actively engaged in research at the agency and bring a significant amount of clinical and scientific expertise to the review process. From the CDRH perspective, Malin noted that formal and informal collaboration is key to understanding needs and opportunities for products. Solving the needs for a certain product may be translated to other types of products as well. The speakers highlighted the numerous interactions among their centers, from sharing supercomputing capabilities to collaborating on the development of modeling approaches, software, and analytic tools.

With regard to dissemination of information, panelists noted that information about a product or process that is generalizable knowledge is released by FDA in the form of guidance documents. The panelists noted that FDA scientists are encouraged to publish their research in the peer-reviewed literature. There is also a wealth of information on the FDA website; for example, the CDER Office of In Vitro Diagnostics has several databases on its website with information about approved and cleared devices.

Professional development for FDA scientists and reviewers is also important to keep agency science on the cutting edge. For example, CDER has weekly "scientific rounds" where novel scientific and regulatory issues are discussed. There are online programs as well as classroom-based programs. Reviewers also need to be able to attend scientific meetings and have access to the latest information.

With regard to potential indicators or metrics of success of the regulatory science initiatives underway, it was noted that it is difficult to measure precisely the public health impact of any particular initiative. While the long-term, big-picture goal is an increased number of approved

MCMs, the metric cannot simply be approvals, because not all products will (or should be) approved. The FDA panelists suggested that metrics should be associated with smaller, incremental steps, such as solving a problem that allows for a potency assay in vitro instead of in hundreds of animals, thereby increasing speed and decreasing the cost of that potency assay.

ENTERPRISE PARTNER AND STAKEHOLDER PERSPECTIVES

Immediately following the panel presentations from FDA representatives, stakeholders representing the other key components of the MCM enterprise provided remarks in which they identified key issues in MCM development and utilization that can be addressed through regulatory science and offered suggestions of regulatory science needs or priorities to advance MCM development.

Department of Defense (DoD)

Gerald Parker of DoD said there is a need for affordable, easy-to-use, rapid, point-of-need diagnostics that can be made available on a global basis and that can be connected to an information backbone so the resulting data can be rapidly shared (within minutes or hours, rather than weeks). This includes diagnostics not only for pathogen identification but also for antibiotic/antiviral resistance patterns, presymptomatic biomarkers, and host response markers.

Biodefense research at the DoD is focused on both traditional threats and endemic diseases that the nation's adversaries could choose to use against U.S. forces around the world. The DoD is working in a collaborative manner, seeking to use platform technologies that incorporate rapid pathogen characterization and the ability rapidly to turn that information into a discoverable product. While the DoD program has a sound science and technology base, Parker noted that the department lacks the ability to rapidly develop discoveries and manufacture new candidates against unknown threats.

The DoD MCM Initiative strategy consists of two major elements, each with multiple initiatives: (1) science and technology (novel platform/expression systems for MCMs, regulatory science technologies, manufacturing technologies for biologics that support good laboratory practices [GLP]/good manufacturing practices [GMP]) and (2) advanced development (further maturation of novel platform/expression systems and integration into a production process; innovative, flexible, and agile manufacturing capabilities).

In closing, Parker said that MCM development needs a clearly defined regulatory pathway for products approved under the Animal Rule, including early and ongoing real-time engagement of all partners. He also noted that a DoD diagnostics leadership meeting held in October 2010 called for more inclusion in discussions and greater collaboration to develop diagnostics and to inform the regulatory roadmap for next-generation diagnostics.

National Institute of Allergy and Infectious Diseases (NIAID)

Michael Kurilla, of NIAID, explained that NIAID is taking a comprehensive approach to MCM development, with certain general criteria for vaccines, therapeutics, and diagnostics. By way of example, he noted that in the case of vaccines, it is important to consider alternatives for immunocompromised persons and special populations such as the elderly and children.

There are several unique aspects of bioterrorism agents that add to the challenge of developing MCMs:

- there are limited facilities to conduct studies under appropriate biological safety containment;
- there is limited prior art on fundamental aspects of specific pathogens (e.g., tularemia);
- there is limited human pathogenesis data available (necessary to inform animal model development); and
- there is increased regulation and oversight of bioterrorism agents (e.g., rules addressing the possession, use, and transfer of select agents).

Kurilla defined MCMs as falling into four broad classes: (1) previously licensed MCMs for which the mechanism of action would support efficacy (e.g., a licensed antibiotic); (2) previously licensed MCMs that are being repurposed for a nonintuitive application (e.g., the oncologic drug, Gleevec, which demonstrates activity against smallpox in vitro); (3) MCMs that are currently in development for other clinical indications (e.g., novel anti-infective agents); and (4) MCMs developed solely for a CBRN application (e.g., an anthrax or Ebola vaccine).

Kurilla posed several questions for consideration regarding the evaluation of new science and technologies:

- What should be done in cases of limited or nonexisting human clinical data with which to define an appropriate animal model?
- How can we study species-specific biological agents, those for which there may be no appropriate animal model?

- What criteria define an animal model correlate?
- Can mechanistic efficacy substitute for disease efficacy? (If the mechanism of an intervention is understood, can that be applied across a wide array of different disease spectrums where that mechanism is identified as crucial for a resolution of that disease?)

In discussion, Kurilla pointed out that much of what was discussed by the panels, and many of the major elements needed, are product-independent regulatory science and product-independent tools. However, the traditional regulatory paradigm is regulation in the context of a product, and there is little interaction with the agency in the absence of a specific product. Product developers approach FDA when they are ready to take a product into pivotal efficacy and safety studies, and it is at that late point that some of the development tools, such as the animal models used thus far, begin to be critically reviewed and questioned (e.g., is the species relevant, is the challenge strain appropriate). Development of acceptable tools has always occurred concomitant with product development, and consequently, Kurilla said, developing these components in a product-independent manner, and in the most expeditious and rational manner, is quite challenging.

Biomedical Advanced Research and Development Authority (BARDA)

The 2009 H1N1 influenza pandemic brought to light some of the challenges of responding to a major public health emergency. Richard Hatchett, chief medical officer and deputy director of BARDA, cited the August 2010 report of the President's Council of Advisors on Science and Technology (PCAST) on influenza vaccine technology, which identified two response issues directly involving regulatory science: the need to improve sterility testing of influenza vaccine, and the need for new techniques to test potency of vaccine preparations (PCAST, 2010). BARDA is investing in research in these areas, he noted.

With regard to regulatory science more broadly, Hatchett said that BARDA is looking to FDA for the following:

- *Internal competency*—FDA needs to have the expertise to keep up with advances in science, to be able to engage creatively with MCM product developers, and to adapt to new technologies (e.g., nanotechnology, bioinformatics, regenerative medicine, in vivo imaging, new approaches to clinical trial design).
- *Capability*—FDA needs to have the capability to help BARDA understand the requirements for proving safety, efficacy, sterility,

and potency of vaccines or other products. This requires FDA to interact with BARDA partners in a collaborative fashion (which the agency is already doing, Hatchett noted).
- *Clarity*—FDA needs to be confident and assertive in defining the requirements for licensure and approval of new products. BARDA is looking for clear pathways, where those pathways can be defined in advance.

Hatchett also emphasized the need to better understand animal models and apply them in a variety of settings.

Hatchett concluded by drawing attention to a forthcoming BARDA request for proposals (RFP's) on multiproduct facilities and rapid response manufacturing capabilities. This will require new approaches from FDA, he said, to be able to license products manufactured in facilities where there may be rapid changeover in response to emerging novel threats (e.g., pandemic influenza, sudden acute respiratory syndrome [SARS]).

Centers for Disease Control and Prevention (CDC)

The Laboratory Response Network (LRN) at the Centers for Disease Control and Prevention (CDC), a key national stakeholder in the MCM enterprise, is currently facing a variety of challenges in the area of diagnostics development, said May Chu, director of the Laboratory Science Policy and Practice Program at CDC. Chu described CDC's most significant current challenge as obtaining FDA clearance of LRN-developed assays (including 11 polymerase chain reaction [PCR]-based assays and seven other assays). Chu noted that an LRN technical review committee oversees assay design, development, validation, and quality assurance prior to deployment, and she anticipated that transformational changes in regulatory science could provide relief while preserving quality and resilience.

Chu also noted that changes to a diagnostic platform require renewed validation. Chu suggested that collaborative discussions are needed to determine what validation is needed when changes occur. A software change, for example, should not necessarily lead to a full reevaluation.

Chu listed the top regulatory science priorities in the diagnostics field as follows:

- Allow for quick and nimble preparedness and response to public health emergencies.
  - Prevalidate and preposition diagnostic tests in the field.
  - Rapid, real-time, step-in-step MCM development with regulatory oversight. This, Chu explained, would allow data collected during an emergency to be used later to validate test platforms.

  - Validation methodology for assessing lot-to-lot differences of commercial products.
- Allow for recognition of the diversity of diagnostic test producers.
- Maintain evidence-based quality and postmarket monitoring with stipulated controls and restricted use.

Academia

A perspective from academia was provided by Rick Lyons, director of the Infectious Disease Research Center at Colorado State University. Academicians are now accepting that for maximal benefit, an innovation must be translated to an application, Lyons noted. Regulatory science targets the pathways that are required for this translation (e.g., biomarkers, animal models, correlates of protection).

The most significant challenge, Lyons said, is the extrapolation of animal immunological and pathophysiological data to the human setting. What makes a good animal model is highly dependent on the research question, he said. One must consider, for example, whether there is similar pathophysiology as the human disease or similar mediators of immune protection. For product development, are there well-defined generic animal models or platforms that could be used? Most of the time, Lyons said, researchers are working with a nonvalidated surrogate that is "reasonably likely" to predict clinical efficacy.

Lyons opined that it is unlikely one species model will reflect human disease adequately and suggested a compartmentalization strategy, pooling data from several species models. This systems and pathways approach would require strong comparative immunology and physiology, he noted.

It will be important to educate the public regarding advances in regulatory science, Lyons pointed out, particularly the use of animal models for approval of products. It has been difficult, for example, to convince people to be vaccinated, or to have their children vaccinated, with products that have been FDA approved based on clinical trials in humans. How much more difficult will it be, Lyons asked, to get them to take a vaccine that has not been tested in humans?

Industry

A biotechnology industry perspective on regulatory science was provided by Eric Rose of SIGA, Inc. The biotechnology industry has developed and manufactures essentially all of the new CBRN countermeasures that have been procured into the SNS, Rose said. Companies are BARDA partners for most of the advanced development contracts. Most companies, however, are small and not profitable, sustained by private capital and government grants and contracts. Biotechnology companies are, Rose stated, "an essential effector limb of the PHEMCE implementation plan."

The goals of regulatory science, Rose said, should be development of a broad array of safe and effective MCMs; alignment of stakeholders, products, and product uses; appropriate transparency throughout the process; and speed (i.e., there should be a sense of urgency as these agents are not just causes of illness, they are potential weapons of mass destruction).

In industry, there is a science to process improvement that Rose suggested can be applied to the process of MCM development. The first step is to design a process (hypothesize). That process is tested by use (experiment), assessed (analyzing performance metrics), redesigned (refine hypothesis), and the cycle continues. The ultimate validation of animal efficacy models requires clinical trials during an outbreak situation. This is something that should be planned for, he said. Rose also noted that, with regard to development of animal models, criteria for euthanizing animals when they have reached a certain degree of illness needs to be transparent and prespecified.

Rose concluded by noting that while robust evidence of meaningful outcomes from controlled clinical trials remains the scientific gold standard for efficacy, the challenges facing MCM development are not necessarily new, and FDA has helped foster other industry segments in the absence of feasible trials, including, for example, orphan drugs, complex medical devices (e.g., artificial hearts), and diagnostics.

Discussion

There was some discussion around reverse engineering to evaluate what processes might work best. One suggestion was to select several products that are currently in the SNS and retrospectively simulate the process as if they had been new molecular entities. It was also noted that there is a lot of animal data available that is associated with approved products. Although these products are not MCMs, it might be helpful to look back at the animal data submitted for product approval and evaluate which models were predictive and which were less so.

Participants also discussed benefit-risk assessments for MCMs and whether the criteria should be the same or different from that for routine products. George Korch of HHS noted that a challenge is conducting a benefit-risk calculus that captures rare and yet highly consequential events. Hatchett added that it is very hard to define, in advance of a real event, criteria for a benefit-risk analysis that are able to take into account the operating environment that exists once the event has occurred. Calculating the risk and benefit of an anthrax antitoxin today, for example, is very different from calculating the risk and benefit of the anthrax antitoxin once there has been a widespread anthrax release. Jesse Goodman of FDA said that the language around FDA's EUA authority regarding the known and potential benefits outweighing the known and potential risks leaves a lot of room for different interpretations. Goodman, Parker, and others all noted that defining the emergency scenario up front would allow for different benefit-risk decisions than those that would be reached for use of common products by generally healthy people. Hatchett cautioned that the calculus done in anticipation of an emergency tends to be much more stringent than that which would actually be made during a real event. To aid benefit-risk decisions, Rose suggested, reviewers of MCMs ought to have the requisite security clearances to be allowed to read the associated confidential population threat assessments. In later discussion about safety assessment, Richard Forshee of the Office of Biostatistics and Epidemiology at CBER, mentioned current agency efforts to develop risk assessment models to support regulatory decision making. These probabilistic quantitative computer simulation models can help explore how different regulatory science options could ultimately have an impact on public health, and thereby improve decision making. FDA is also engaged in a number of computational toxicology computer simulations to help assess, for example, potential risk from vaccine adjuvants.

Key Messages: Enterprise Stakeholder Perspectives

- The regulatory paradigm for MCM development needs to be supported by new regulatory science and evaluative tools that are product-independent. There is a need for a format to permit engagement between product developers and FDA outside the context of a specific product approval.
- MCM development needs more clearly defined regulatory pathways. Priorities include:
    - Products approved under the Animal Rule, and
    - Diagnostics—prevalidated and pre-positioned in the field
- The benefit-risk calculus may be different for MCMs to be used in low-probability/high-consequence events than for traditional products.
- Repurposing of previously licensed products needs to be studied in a systematic and comprehensive manner.

3

Cutting-Edge Efforts to Advance MCM Regulatory Science

NONCLINICAL APPROACHES TO ASSESSING EFFICACY

A challenge facing developers of MCMs is how to increase the predictive value of nonclinical data, said panel moderator Lauren Black, senior scientific advisor at Charles River Laboratories. In the absence of clinical trials, nonclinical data can, for example, help define a human dose regimen and predict a reasonable likelihood of human efficacy. In addition to animal models, other nonclinical tools such as *in silico* biology and biomarkers can be employed to inform and advance MCM development.

In Silico Approaches to Efficacy Assessment of MCMs

A systems biology approach to health and disease acknowledges that there are likely complex molecular mechanisms, with groups of molecules, genes, proteins, and metabolites working in a coordinated fashion, that differ in healthy versus diseased states, explained Ramon Felciano, founder of Ingenuity Systems. These molecular mechanisms trigger higher-order cellular mechanisms and disease mechanisms that drive overall physiology (Figure 3-1). Technologies that have emerged over the past decade or so (e.g., genomics, proteomics, metabolomics) have generated a flood of new data, driving the need for new types of analytics such as *in silico* or computer modeling of biology. These new data enhance and align with existing knowledge of disease pathways and mechanisms from the literature.

A typical systems biology approach is philosophically data driven

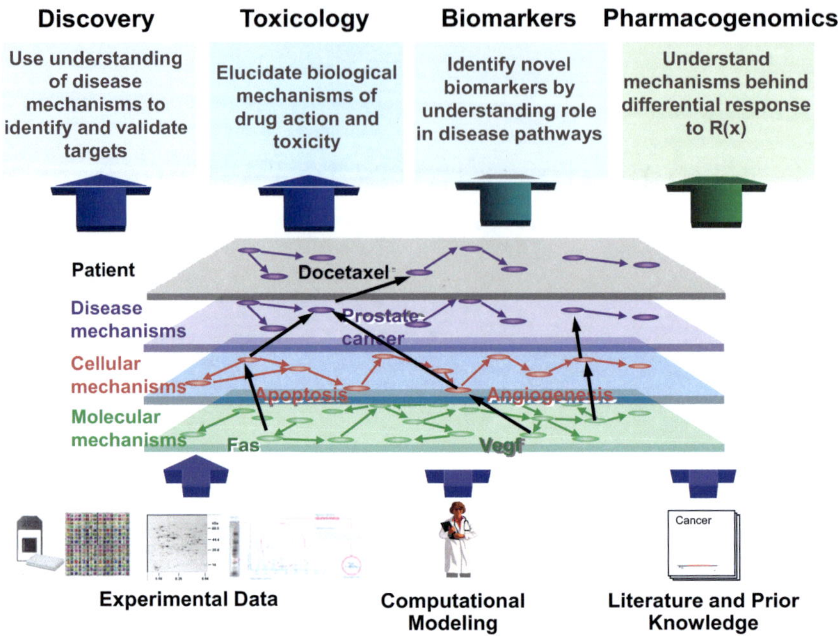

FIGURE 3-1 *In silico* modeling of disease mechanisms for drug development. SOURCE: Ramon Felciano. 2011. Presentation at IOM workshop; Advancing Regulatory Science for Medical Countermeasure Development.

and holistic, Felciano said, using computer-based tools and techniques to model and understand complex biological function. Experimental designs are typically comparative in nature (e.g., healthy versus disease, disease versus treatment, dose response). The complexity and volume of the data that are generated by these approaches typically require fairly sophisticated computational and statistical modeling for analysis and prediction. Research teams are often interdisciplinary by necessity, with therapeutic area researchers, computer scientists, statisticians, and others working together.

Primary benefits of this approach, Felciano said, include better understanding of disease progression, generation of novel hypotheses for therapeutic or diagnostic targets (i.e., biomarkers), and characterization of plausible mechanisms that correlate with these diagnostic and prognostic markers.

Compared with other therapeutic areas, there has been relatively little research done in the area of MCMs using a systems biology approach, Felciano noted. He cited one retrospective study of yellow fever vaccine that demonstrates the potential of *in silico* approaches to modeling. Querec and colleagues (2009) used a systems biology approach to identify early gene signatures that predicted immune response in humans to the yellow fever vaccine.

There are several challenges to using *in silico* techniques in MCM efficacy studies, Felciano said. As this is a new field, no dominant modeling formalisms have yet emerged, and there is a lot of new math being generated alongside the new data. Some of the "-omics" technologies are still relatively new, and there are issues to be addressed, for example: measurement accuracy and reproducibility, false positive results, and cost effectiveness. Felciano noted that FDA has a Voluntary Exploratory Data Submission (VXDS) program in which industry submits candidate datasets that FDA can use to evaluate the regulatory applicability of these new approaches. Other challenges are that systems biology experiments are complex and interdisciplinary, requiring substantial time, interdisciplinary expertise, and resources for analysis. Thus far, there are few successful applications of *in silico* techniques to infectious disease. In addition, there are few good predictive models to bridge animal data to humans.

To leverage *in silico* modeling for MCM development, Felciano said there is a need for more VXDS submissions for clinical infectious disease and MCM studies, with an emphasis on proposals including genomic markers of efficacy. Secondly, Felciano recommended a public "genome-to-phenome" database that characterizes, at a systems biology level, how existing animal models are representative of given target endpoints and underlying mechanisms. This would allow for assessment of concordance between existing "well-characterized animal models" and *in silico* links between molecular systems and animal study endpoints. Finally, Felciano recommended the collection and integration of quantitative data on human and animal model immunity in normal, vaccinated, and infected individuals. This would allow for analysis of efficacy and common markers of response for existing treatments between humans and animals, as well as between animal models.

In summary discussion, participants discussed clinical trial simulations, in which virtual patients are put through *in silico* trials, a process that allows a company to model a wide range of trial designs and analysis methods, with the goal of reaching programmatic decisions more quickly, more cheaply, and with greater certainty. To leverage *in silico* modeling, there is a need for more complete, shared databases of human and animal study data (e.g., genome-to-phenome information; data on human and animal model immunity in normal, vaccinated, and infected individuals).

It was also suggested that a Bayesian, model-based, predictive framework should be developed that would essentially create *in silico* animals and a virtual human. It was noted that such a project would require permissions, funding, and collaborations on the scale of IBM's Watson project or the Manhattan Project.

Using Biomarkers to Connect Animal Systems with Clinical Efficacy

Measurements of biomarker molecules are intended to allow connection of physiological changes with changes in outcomes or risks, explained Leigh Anderson, founder and CEO of the Plasma Proteome Institute. Biomarkers measured in blood and tissue are generally proteins, measured by immunoassay, or mRNAs (ribonucleic acid), measured using microarrays. Anderson offered cardiac troponin as an example of a successful biomarker; an increase in this protein is indicative of a recent heart attack.

Candidate biomarkers can be identified via *in silico* modeling studies, experimental studies, and by analogy with other species. However, Anderson noted, establishing biomarker validity requires significant effort, and all methods of hypothesizing biomarkers are extremely failure prone (> 99 percent attrition).

There are 109 proteins for which tests have been approved or cleared by FDA, and 96 additional proteins that can be tested for using laboratory-developed tests (that have not been reviewed by FDA). Approval of new protein biomarkers occurs at a very slow rate, Anderson noted: about 1.5 new protein biomarkers per year over the last 15 years. This rate, he said, is insufficient to meet broad clinical needs, without even considering MCM development.

Part of this dearth of new biomarkers is caused by the lack of a real pipeline for systematic discovery, development, and marketing of biomarkers. There are technological issues, including the lack of reliable, high-throughput assays for most biomarker candidates and the slow pace of development of new protein assays. There are also challenges in accessing large, existing sample sets in which to test the clinical relevance of a biomarker. The basic understanding of the mechanism for cross-species extrapolation is also very poor.

An ideal biomarker measurement method would provide certainty as to analyte structure, Anderson said. It would include internal standards and would have a method of eliminating interferences. He noted that mass spectrometry allows for very high-specificity measurements of proteins, with quantitative accuracy and internal standards, and it is inherently multiplexable. These assays can be developed very quickly, and there is the potential that the improved science content could allow more rapid approval by FDA.

In conclusion, Anderson said the challenges of identifying biomarkers for development of MCMs are similar to those for biomarkers for general clinical use. He emphasized the fundamental need to develop a biomarker pipeline capable of systematically addressing complex biology. Biomarkers of efficacy for MCMs must be established in advance for the species involved in MCM testing. This requires a systematic evaluation of candidate biomarker homologs across a range of species, something that has not been done thus far. Success of an MCM biomarker also relies heavily on parallel mechanisms of disease, treatment efficacy, and recovery across species. Anderson also recommended that biomarker measurement technology be based on a high-confidence, rapidly approvable analytical method.

It is now feasible, Anderson said, to make specific, FDA-approvable assays for all 20,000 human proteins. Despite statistical design challenges, it is near feasible, he suggested, to test all possible proteins as candidate biomarkers against a broad range of diseases. If this were done, it could establish broad parallelism between human and animal systems.

Marietta Anthony of the Critical Path Institute (C-Path) presented information about efforts of the Predictive Safety Testing Consortium (PSTC) to achieve FDA qualification of seven renal toxicity biomarkers. The specific context of use that FDA allowed was for drug induced kidney injury in GLP rat studies and to support clinical trials. She remarked that the next step for C-Path is to conduct human clinical studies to assess their seven renal biomarkers. If the data are found to be important, they will be submitted to FDA for qualification. Donna Mendrick of the National Center for Toxicological Research (NCTR) at FDA commented that translating biomarkers can be extremely challenging. Mendrick noted that for kidney biomarkers, the gold standard in animal studies is histopathology, while in the clinic, the gold standard is measurement of serum blood urea nitrogen (BUN) and creatinine, which become abnormal at a later stage in disease. Anthony noted that the seven renal biomarkers that were qualified reflect histopathology far more effectively than BUN. Vikram Patel of the Office of Testing and Research at FDA's CDER reminded participants the ultimate proof of efficacy of an MCM only comes when it is used in humans. In that regard, having a biomarker is very important to help assess whether the MCM, or which of several MCMs, is effective in an emergency situation. He expressed concern that very little attention is being paid to development of biomarkers.

Animal Models of MCM Efficacy

Throughout the workshop a number of participants discussed limitations of animal models.

Michael Kurilla of NIAID set the stage by noting in his remarks in the session on enterprise stakeholder perspectives (Chapter 1) that animal models are critical to MCM development; however, most animal models are not suitable for a number of potential reasons. Animal models are infection models, Kurilla reminded participants, not disease models, and some infectious diseases are uniquely human diseases (i.e., there may not be any appropriate animal model). In addition, pathogenesis differs among various species, animals may not fully model host defense responses, and the availability of species-specific reagents may preclude the ability to define correlates. Extensive pathogenesis and natural history studies are necessary to demonstrate the validity of a particular species to replicate a human disease. There are also feasibility issues with conducting pivotal efficacy studies in animal models, including the development of GLP animal models to support licensure.

Elizabeth Leffel of PharmAthene provided formal remarks about animal models, and a panel discussion ensued. In developing MCMs under the Animal Rule, stakeholders need to think of animal studies as the equivalent of traditional phase I to II clinical trials, said Leffel. Leffel emphasized that while aspects of animal models can be standardized, animals cannot be "validated," just as we cannot validate humans in clinical trials. She also noted that both humans and animals are heterogeneous populations, and no model can be 100 percent predictive of what will happen in humans.

The primary regulatory science tool for animal models is, of course, the FDA Animal Rule. There is a relatively new draft guidance published to support the Animal Rule, entitled "Qualification Process for Drug Development Tools." This guidance, Leffel clarified, is not a mechanism to discuss product-specific tools or assays; rather, it is meant to address how animal models can be applied broadly to more than one drug.

Leffel identified four key regulatory science needs relative to animal models. First, she said, the essence of the Animal Rule needs to be consistently defined to product sponsors. There are different interpretations across FDA divisions, she noted, and sometimes between reviewers within the same division, of how to apply the Animal Rule. Second, appropriate review of MCMs based on risk and benefit is needed. These are high-risk, life-threatening diseases, about which clinical knowledge is often limited. Third, Leffel noted, there is a need for precompetitive mechanisms to share basic animal model information quickly. This includes shared proof-of-concept studies to avoid duplication (e.g., for NIAID-sponsored studies, information on basic models for vaccine studies is available in cross-referenced master files for sponsors). Fourth, as noted by others, there must be ways to bridge nonclinical models to expected human outcomes, such as surrogate markers, correlates of protection, clinical observation in animals, and pathology.

Moving forward, the first priority, Leffel said, is to develop a strategic plan for applying the Animal Rule. She suggested:

- This includes finalizing the draft guidances to reflect current FDA thinking[1] and then applying these standards consistently within and across divisions at FDA and across sponsors. Areas that could be standardized by disease should be identified, and those areas that cannot be should be recognized. The strategy should also include preparing the MCM enterprise to accept more risk, as well as adopting provisions to mitigate risk (by, for example, special licensing conditions such as restricted or conditional licenses).
- A second priority, Leffel said, should be to leverage existing initiatives or form new partnerships to enhance data sharing. There are a lot of partnerships already in existence, she noted, and we need to start using them more effectively. She cited the FDA-NIH regulatory science initiative as a potential opportunity to allow FDA to leverage scientific resources from NIH and further engage FDA scientists in professional development.
- Third, she suggested, licensure review could be expedited by engaging cross-functional expert teams early on. Specifically, Leffel noted, in addition to meetings between product sponsors and FDA, it might improve communication further if an FDA scientist could also be present at the regular meetings between product sponsors that have U.S. government contracts and the relevant funding agency, at least at significant time points or milestones.
- Public-private partnerships, such as early development partnerships between industry and DoD and NIH labs, could be effective, and cross-industry precompetitive collaboration models should be pursued. Leffel also suggested that the agency should initiate a risk communication strategy to the public and establish dedicated cross-divisional review teams to evaluate MCMs under the Animal Rule.

Animal Model Case Study and Discussion

Drusilla Burns from the Office of Vaccines Research and Review in FDA's CBER offered as a case study the pathway to licensure for anthrax vaccines. Animal models were developed, she said, that were thought to be appropriately reflective of human disease. It was demonstrated that an immune marker, anthrax toxin neutralization antibodies, correlated with protection in the animals, and the protective level of antibody was

---

[1] A participant from FDA clarified the status of the draft Animal Rule guidance. Following the public comment meeting in November 2010, the draft guidance is undergoing major revisions and, as such, will not be finalized but will be republished as a draft to allow for another comment period on the revised guidance.

identified. Further studies demonstrated that the assay that is used to measure these antibodies was species independent, allowing for bridging to humans (i.e., in a clinical trial, measuring the antibody levels in humans could be used to predict the potential efficacy in humans). While this may sound simple, Burns said, it was very resource intensive, involving convening a workshop, conducting a literature review and interviews with experts, and forming an interagency animal studies working group that called upon vaccine manufacturers, academicians, and government contractors as needed. She emphasized the importance of the scientific partnerships between FDA scientists and other government scientists or outside scientists, and the involvement of diverse disciplines.

Judy Hewitt, chief of the Biodefense Research Resources Section at NIAID, emphasized the importance of qualification of animal models in a product-neutral manner. Patel of FDA suggested having a control animal dataset in a national database to which sponsors could compare their animal test data. Leffel commented that organizations such as the Alliance for Biosecurity, a public-private partnership, have taken steps to pursue development of a shared database of anthrax animal model data; unfortunately, that effort was underfunded. She noted that BARDA has picked up some of this work in anthrax and is in the early stages of working with industry partners to conduct meta-analyses on contributed data. She emphasized that adequate funding is critical to the success of these types of initiatives.

In summary discussion, it was noted that there is a clear need for a better understanding of animal models and how to apply them in a variety of settings. One of the most significant challenges is the extrapolation of animal immunological and pathophysiological data to the human setting, and participants discussed the need for new approaches to bridge nonclinical models to expected human outcomes (e.g., surrogate markers, correlates of protection). A number of workshop participants noted that it is unlikely one species model will reflect human disease adequately, and a compartmentalization strategy, pooling data from several species models, was proposed. Workshop co-chair, Les Benet, Professor in the Department of Biopharmaceutical Sciences of the University of California, San Francisco, called attention to a series of five forthcoming papers, part of the P*h*RMA initiative on predicting models of efficacy, safety, and compound properties, that found that, for 108 new molecular entities where both human PK and animal data were available, the animal models were poor in predicting (Poulin et al., 2011a,b; Ring et al., 2011). There was also interest in setting up precompetitive mechanisms to share basic animal model information quickly (including proof of concept studies to avoid duplication).

Picking up on earlier discussions, Benet suggested that a retrospective look at historical animal data from approved vaccines, anti-infectives, and other products could help inform discussions about the Animal Rule. He proposed looking at the data from animal studies as if that were all that was available, and making a hypothetical approval decision under the Animal Rule criteria, and then comparing how well that correlates with what is known from the human clinical trials that the actual product approval was based on. In other words, asking "Using all of the predictive methodologies that we have available today, if we approved this product under Animal Rule, would we have made the 'right' decision?"

In discussion about this proposal, Robert M. Nelson, senior pediatric ethicist at FDA, noted a concern that most animal work is done for preclinical toxicology purposes, and there may not be a robust enough dataset around the appropriate animal model for this type of exercise. A participant from industry countered that they often conduct proof-of-concept efficacy studies in mice and rats prior to initiating phase II trials in humans. Ed Cox, Director of the Office of Antimicrobial Products within the Office of New Drugs of CDER, said to keep in mind there are different types of animal models, those intended to look at an activity (e.g., pharmacokinetic/pharmacodynamic [PK/PD]) and those that are intended to mirror the human condition (involving an actual tissue site where infection would occur and some of the local factors at that site). In addition, there are models of infection and models of disease. Participants also noted the challenge and the importance of comparing "apples to apples" when looking at historical data. Adding to the complexity is the fact that tests are done by different laboratories with different standards. Another participant suggested that an alternative approach could be to conduct a new animal study with a current, approved drug or vaccine, in an appropriate model, and base the predictions on that data.

Key Messages: Nonclinical Approaches to Assessing Efficacy

***In Silico* Approaches and Biomarkers**

- Clinical trial simulations hold promise for modeling a wide range of trial designs and analysis methods and could facilitate reaching programmatic decisions more quickly, more cheaply, and with greater certainty.
- There is a need for a biomarker pipeline capable of systematically addressing complex biology. Efforts should include systematic evaluation of candidate biomarker homologs across a range of species.
- "Big science" could be envisioned for new projects, such as:
  - A Bayesian, model-based, predictive framework could be applied to create a "virtual human"; such a project would require momentum and collaboration on a large scale.

continued

Key Messages Continued

- Make specific assays for all 20,000 human proteins; statistical design challenges would need to be overcome.

Animal Models

- Building databases of existing animal models (genome to phenome) could allow for assessment of concordance between existing "well-characterized animal models" and *in silico* links between molecular systems and animal study endpoints.
- A control animal dataset in a national database would permit comparisons by sponsors of their animal test data.
- Scientific partnerships, including creation of an "ecosystem" of collaboration and a multidisciplinary approach, is important for addressing difficult regulatory science problems in assessing efficacy.
- Funding and substantial resources are essential to sustain interagency, public-private, and other enterprise partnerships and collaborations.

SAFETY AND REAL-TIME MONITORING

In a public health emergency, some of the MCMs used may be new molecular entities for which efficacy studies in humans were not done, and predeployment safety information is limited, said panel moderator, Carl Peck, of the University of California, San Francisco. He noted that once a new MCM is deployed, it will be especially important to monitor for side effects and to confirm effectiveness (so that use of an MCM that is not effective can be discontinued and further risk of adverse events reduced).

Toxicology Markers

Robert House, president of DynPort Vaccine Company, presented about toxicology markers from a vaccine development standpoint, noting that there are a variety of primary toxicological concerns. Local toxicity or "reactogenicity," while not a main concern for small molecules, is a primary concern in developing vaccines. As with any drug, one must also be concerned with systemic toxicity. Toxicity testing is performed under GLP conditions to ensure the cleanest results, using GMP (or GMP-like) material, in a relevant animal model, House said. For vaccines, a standard toxicology profile must also include assessment of immunogenicity. Developmental toxicity and immunotoxicity are also assessed. Vaccine adjuvants must be tested as if they were a new chemical entity (and as such are tested twice, alone and as part of the vaccine). Other additives

TABLE 3-1 Prediction of Clinical Outcomes: Preclinical Toxicology Studies vs. Clinical Studies

Parameter	Toxicology Studies	Clinical Studies
Survival	Yes	Yes
Pain upon injection	Difficult to assess	Yes
Fever	Dependent on animal model	Yes
Headache/malaise	No good animal models exist	Yes
Injection site reactions	Yes	Yes
Clinical signs	Yes	Yes
Body weights	Yes	Useful?
Clinical pathology	Yes	Yes/not usually
Necropsy, histopathology	Yes	Generally not
Antibodies	Yes	Yes
Immunotoxicity	Dependent on animal model	Yes/not usually done

SOURCE: Robert House. 2011. Presentation at IOM workshop; Advancing Regulatory Science for Medical Countermeasure Development.

that go into vaccines, such as excipients or preservatives, must also be individually assessed for toxicity. Depending on how a vaccine is administered, it may also be necessary to assess the toxicology of the administration device.

Standard preclinical toxicological endpoints include body weights (as a measure of robust health); clinical observations (are the animals behaving normally); clinical pathology (including hematology, clinical chemistry, and other immunogenicity studies); anatomic pathology (including organ weights and histopathology to assess intended effect at the immune system target, as well as any effects at other points in the immune system); and local tolerance.

House compared preclinical toxicology studies to clinical studies in their ability to predict clinical outcomes (Table 3-1). He noted that several parameters (in italics)—pain upon injection, fever, headache and malaise, and injection site reactions—are often considered to be rather subjective and can be difficult assess in animal models.

Electronic Monitoring of Adverse Events

Kenneth Mandl of the Harvard Medical School Center for Biomedical Informatics characterized four main sources of clinical electronic health data:

1. Reported data—Voluntary or mandated reporting of adverse events to FDA or disease outbreaks to CDC;
2. Ambient data from the health system—Data that are produced through the routine processes of care;
3. Meticulously collected data—Registries on a selected population of patients; and
4. Patient-reported data—Consumer technologies, social networks, personal health records, and other technologies that can be accessed directly.

Mandl cited a study by Basch (2010) that highlights the value of patient-reported data. Individual contributions to drug safety data can include adverse effects, efficacy endpoints, adherence, satisfaction, and quality of life, as well as concomitant over-the-counter and complementary/alternative medicines.

A challenge to electronic data monitoring is that electronic medical records are managed by software that runs locally. Each one of these systems is different and often proprietary and unmodifiable. "The data are hard to get in and virtually impossible to get out," Mandl said.

To help address this, Mandl and Kohane (2009) have proposed a platform approach to health data software design, for which many applications or "apps" could be developed (similar to the iPhone platform, Mandl explained). He noted that in such an environment, apps compete with each other in the apps store, so functionality and usability would be expected to improve continually and prices would be competitive as well. Mandl also referred workshop attendees to the "SMART Apps for Health" challenge on the challenge.gov website, a contest to develop apps that provide value to patients.

Mandl noted there has been over a decade of work on biosurveillance (real-time monitoring for detecting an emerging epidemic or an outbreak), and there are some fairly sophisticated systems and techniques that he suggested could be relevant to real-time safety surveillance of MCMs.

In closing, Mandl said that the Obama administration has made a $48 billion investment in health information technology for use across multiple sites, predominantly primary care facilities and hospitals, and he encouraged FDA to become very involved in the discussions of health information technology deployment.

Discussion

Richard Forshee of CBER explained that voluntary reporting of data (e.g., to the Vaccine Adverse Event Reporting System or VAERS) is a key component of FDA's activities in near real-time safety surveillance in the postmarket environment. An advantage of this passive reporting system, he said, is that it is fast. However, as it can be very labor intensive to sort through the many records in the system, the agency is developing text-mining systems that can narrow down the data to sets of adverse event reports that are most relevant for the medical officers to address.

Henry Francis, Deputy Office Director in the Office of Surveillance and Epidemiology of CDER, said that new approaches are needed in data collection (as noted, current systems are passive), quantitative analysis (FDA receives roughly 800,000 reports every year that are read by 43 people), qualitative assessment (e.g., a pharmacist may report that people are taking a medication incorrectly, such as chewing a Spiriva capsule instead of inhaling the contents), and reporting (to regulatory decision makers and the public).

Robert C. Nelson of Product Safety Assurance, Inc., recommended that the Postlicensure Rapid Immunization Safety Monitoring (PRISM) system that was developed to monitor the safety of the 2009 H1N1 influenza vaccine be institutionalized and used for all vaccines. He also noted that the FDA Adverse Event Reporting System (AERS) could be a starting point to help detect serious rare events associated with MCMs; however, monitoring the efficacy of the nonvaccine MCMs will be a significant challenge.

A participant drew attention to a collaboration between the Indiana University (IU) Medical Center, Eli Lilly, and the Regenstrief Institute for real-time adverse event monitoring. Data are collected at the point of care via a MedWatch-like form embedded into the electronic medical record system at IU. If a physician wants to report an adverse event, the form pops up in the system, and the completed form is simultaneously entered into the patient's electronic medical record and forwarded to the manufacturer if there is a drug involved.

Mendrick cautioned that it can be very difficult to discriminate between drug-related postmarket adverse events and disease-related processes in humans.

Key Messages: Safety and Real-Time Monitoring

- Collaborative working groups should be convened to share data and experiences with respect to safety biomarkers.
- "App" technology for surveillance, response, and adverse event monitoring holds promise.

ANTIMICROBIALS

When assessing microbial threats, strain identity is secondary to the organism's resistance profile, asserted Kevin Judice, CEO of Achaogen, Inc. For example, the gram-positive bacteria, methicillin-resistant *Staphylococcus aureus* (MRSA) and *Bacillus anthracis* or anthrax, are similar at a genetic level and phenotypically. As such, an antimicrobial to treat MRSA will generally be effective against anthrax as well. If the microbial strain in question is, for example, susceptible to ciprofloxacin, there is generally no problem with treatment. However, strains that are resistant are essentially untreatable. The rise of ciprofloxicin resistance in *E. coli* over the past decade has been significant. *Yersinia pestis* (plague) is closely related to *E. coli,* and resistance genes can transfer naturally, resulting in ciprofloxacin-resistant *Y. pestis*. So although this is a naturally occurring process in "civilian strains" (those found in nature), the result is a resistant strain of a biothreat agent.

Strategies to address drug-resistant civilian strains can suffice for MCMs as well, Judice said. The early development pathways are essentially identical (excluding any specialized microbiology, or specialized laboratories for handling threat strains). As later-stage clinical trials with threat agents are impossible, it is necessary to rely on civilian indications for clinical trials.

As an example, Judice discussed nosocomial pneumonia, which is a good proxy for plague. Judice provided background that an investigational Achaogen compound, ACHN-978, has been shown to be active against resistant strains of pseudomonas and acinetobacter, as well as *Y. pestis* and *Francisella tularensis*. To demonstrate the clinical utility of ACHN-978, the company is focusing on nosocomial pneumonia associated with resistant pseudomonas and acinetobacter.

Current guidance on trial design, however, calls for trials that Judice said are often too big or too complex for a small company to conduct. In a standard noninferiority clinical trial design, patients with suspected infection are typically randomized to control (standard of care) or investigational treatment arms before bacterial culture results are available (i.e., before the pathogen or its resistance profile is known). A subset of each group will have resistant pathogens; however, because randomization was done before their resistance profile was known, groups may be unbalanced with regard to patients with multidrug-resistant strains, making analysis challenging. If the intent is to test the investigational product for efficacy against resistant strains, this is not an optimal design. Judice suggested that, under current guidelines, such a trial would need roughly 2,500 patients, would cost around $300 million, and would take five to seven years to complete, which is generally beyond the resources or abilities of a small company.

To move forward, Judice said that industry and FDA must work together, focusing on reassessing benefits and risk. Industry, he said, needs to develop better molecules that solve real problems (e.g., antimicrobial resistance) and demonstrate benefits in the clinic (superiority versus resistant strains). FDA may need to accept more risk, he said, such as alternative noninferiority margins for a novel agent to treat life-threatening multidrug-resistant infections.

Judice emphasized the need to enroll more patients with resistant organisms into clinical trials and suggested a superiority trial design in which patients are enrolled but are not randomized until after culture and sensitivity results are known. To facilitate this, he added, rapid diagnostic tests are needed. Patients with resistant pathogens would then be randomized to the standard of care or investigational therapy.

In closing, Judice expressed concern that we are entering a "post-antibiotic era" for some types of infections. We must work together to solve these problems—better molecules, improved clinical and regulatory pathways—soon, he said.

Discussion

Many participants noted that there is a great need for new antibiotics for patients. Cox expanded on the concept of developing compounds for conventional disease that are also useful against a particular threat agent, often referred to as "dual use" drugs.[2] While this may not be applicable in all settings, he said, studying a drug in a patient population with a common infectious disease provides an opportunity to collect safety, PK, and appropriate dosing data. It also provides an important opportunity to evaluate efficacy; often, for example, the pathology of the conventional disease may have similarities to the disease caused by the threat agent. In addition, if a product has a market for treatment of a conventional disease, it would generally be in production and available during a public health emergency.

Cox concurred that rapid diagnostics could play an important role in making clinical trials of antimicrobial drugs more efficient. Another challenge for antimicrobial clinical trials is that concomitant or prior antibacterial drug therapy cloud the assessment of the efficacy of the investigational drug. Resource barriers, as discussed by Judice, are also a concern,

---

[2] Note that *dual use* is sometimes used to refer to the potential for technology to be both used for biomedical research and misused for hostile purposes (e.g., antibiotic-resistant strains developed to facilitate research on relevant antimicrobial therapies also being used as agents of bioterror). However, in this summary, *dual use* refers exclusively to a product and its potential to treat both conventional and bioterror-related diseases or conditions.

and Cox suggested that economic incentives may be helpful or necessary to offset some of the economic risk that product developers face.

Panel moderator Linda Miller, director of Clinical Microbiology at GSK, suggested the need for guidance that better addresses the issues of antimicrobial trials that were raised (e.g., size of trials, patient populations, appropriate margins for noninferiority trials, use of rapid diagnostics). She also suggested that approval based on pathogen-specific indications be considered.

Workshop co-chair John Rex, of AstraZeneca, raised a concern that the superiority trial approach may work well for the first new drug being compared to older drugs, but it could make it impossible to get further comparable new drugs approved as subsequently superior to that first new drug. Judice responded that a good, well-controlled superiority trial against classic drugs could then set the stage for better noninferiority trials down the road, with potentially more appropriate noninferiority margins. Cox agreed, noting that if a new drug that shows superiority is found to be safe and effective, then it would become the standard of care on which to base a noninferiority margin. It is important to have pathways available for noninferiority trials in conditions such as hospital-acquired pneumonia, ventilator-associated pneumonia, complicated urinary tract infection (UTI), complicated intra-abdominal infections, or community-acquired pneumonia.

A participant pointed out there is a rich trove of data housed at FDA from many years of clinical trials in which there were patients enrolled who had multidrug-resistant organisms and for whom the investigational treatment failed. He suggested there should be a way to de-identify and share those data for reference by researchers.

Key Messages: Antimicrobials

- There is a need to enroll more patients with resistant organisms into clinical trials through superiority trial design, with randomization after sensitivity results are available. Regulatory guidance should be developed to support this, addressing:
  - Trial design issues such as trial size and appropriate margins for noninferiority trials.
  - Use of rapid diagnostic tests.
- Consider making available data currently held at FDA on failure of investigational treatment in patients with multidrug-resistant organisms.

VACCINES AND ADJUVANTS

Vaccines

The current product development timeline for vaccines spans 10 years or longer and can cost $500 million or more, said Alan Magill, program manager at the Defense Advanced Research Projects Agency (DARPA). Vaccine development is also a very-high-risk endeavor.

Knowledge of a gene sequence of a recognized immunogen from a known pathogen (e.g., influenza hemagglutinin) does not guarantee an immunogenic vaccine candidate, Magill pointed out. He commented that vaccine discovery tends to be an empiric trial-and-error process, adding that we need a better understanding of how to design and build an immunogen or antigen that leads to protective antibodies. Toward this goal, the DARPA Protein Design Processes Program is developing tools for the design and synthesis of new functional proteins. Researchers ultimately hope to be able to design, within 24 hours after notification of a threat, a new complex protein that will inactivate the pathogenic organism.

Another issue for vaccine development is establishing immune correlates of protection, which Magill said are really biomarkers for efficacy (e.g., hemagglutinin-inhibition titers for influenza, neutralizing antibody titers for yellow fever). Identification and qualification of these biomarkers starts with collection of specimens and correlation of data to clinical outcomes in clinical trials.

With regard to animal models, the question is whether they are predictive of vaccine protection in humans. Animal models of vaccine protection are expensive and increasingly more difficult to do. Magill questioned the need for GLP toxicology studies in animals prior to phase I clinical vaccine studies, as in his experience they rarely identify problems. DARPA is currently using the Modular IMmune In Vitro Constructs (MIMIC, from Sanofi VaxDesign) system to determine human immune responses directly from an antigen in an assay plate. The ultimate goal would be for this system to eliminate the need for human testing in some settings. In reality, Magill said, it should assist in culling down selection of candidates; clinical trials will likely always be needed. Human tissue engineering is another area that DARPA is aggressively moving into as a potential tool for testing medicines in lieu of animal models (Figure 3-2).

Magill commented that in biological manufacturing, regulatory requirements have led to a common notion that the "process is the product," in that licensure requirements adhere to the "recipe" of how the product is manufactured. Conventional wisdom therefore holds that a product-specific facility is needed, which is very expensive, lengthy, and makes technology transfer particularly challenging. Magill described DARPA efforts to address this by creating "modular GMP" units that can be moved around (Figure 3-3). These mobile bioprocessing facilities

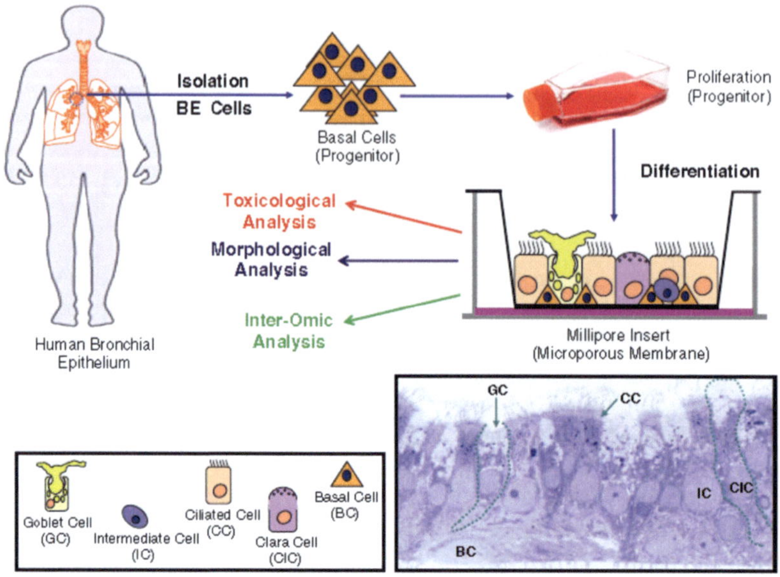

FIGURE 3-2 Human tissue engineering as a potential tool for testing medicines in lieu of animal models.
SOURCE: Alan Magill. 2011. Presentation at IOM workshop; Advancing Regulatory Science for Medical Countermeasure Development.

FIGURE 3-3 Mobile bioprocessing facilities (MBFs).
SOURCE: Alan Magill. 2011. Presentation at IOM workshop; Advancing Regulatory Science for Medical Countermeasure Development.

(MBFs) would have, for example, a protein-processing suite (or whatever is needed) and could result in reduced manufacturing costs and increased flexibility.

Another issue for consideration is antigens versus adjuvants versus vaccines. As subunit vaccines have not proven sufficiently immunogenic, various adjuvants have been used to try to boost immunogenicity and protection. Currently, FDA views approval of a vaccine as approval of a formulation (e.g., antigen and adjuvant). Magill suggested that there is a need for a qualification process for adjuvants, with generation of a drug master file where it would be possible to evaluate multiple antigens.

In closing points, Magill reminded participants that technology will not solve all of the problems of drug development and MCM development in particular. There is a need for better understanding of natural history and pathogenesis of disease and of immunology.

Adjuvants

Debbie Drane, senior vice president for Research and Development at CSL Biotherapies, highlighted some of the main problems faced in adjuvant development. Adjuvants are essentially platform technologies, she said, but they are not viewed that way by regulators (or product developers). Adjuvants themselves are not registered. There is a need to better understand the mechanisms of action of adjuvants. Another unknown is the long-term effects of strong immunopotentiators. There is a lack of preclinical biomarkers particularly for safety of adjuvants. There is also a lack of transparency around adjuvant research, as adjuvant development has largely been industry based.

Drane offered the CSL adjuvant, ISCOMATRIX, as a case study in adjuvant development and evaluation. ISCOMATRIX is a proprietary, saponin-based adjuvant in which ISCOPREP saponin is complexed cholesterol and phospholipids. CSL sought to understand the mode of action (immunogenicity and safety) and the mechanism of action. Animal models were used to evaluate immunogenicity. Different adjuvants have effects on different arms of the immune system, she noted; for example, water-in-oil adjuvants induce antibody responses but are less effective in inducing T-cell responses. Animal models were also used to provide information on the kinetics and potential mechanisms of the response. It can be very useful, Drane said, to understand which immune cells the vaccine formulation is targeting. In vitro studies in human immune cells are done to help link human response to the animal models. Another approach is using human cell lines and in vitro biomarkers (e.g., induction of pro-inflammatory cytokines) to predict potential safety signals associated with adjuvants. Gene profiling can also aid understanding of the tolerability of vaccines.

Safety of an adjuvant is key, Drane emphasized. CSL and other partners are working to establish an integrated database of adjuvant-related safety information from both nonclinical and clinical studies.

In summary, Drane said that vaccine developers must share and evaluate knowledge, and she suggested establishing an adjuvant advisory group that includes FDA, industry, and academia to convene workshops around a variety of adjuvant-specific issues such as safety biomarkers, perceived biothreats, and adjuvant combinations. Drane concurred with Magill that there is a need for a different approach to licensing adjuvants for MCMs, such as broad licensure for adjuvants that could then be used with virtually any MCM vaccine.

Discussion

Hana Golding, chief of the Laboratory of Retrovirus Research at CBER, said that one of the key strengths of FDA scientists is versatility—specifically, the ability of scientists to move from one pathogen to another, from one type of research to another, to address developing diseases. FDA scientists are also in a position to proactively identify gaps that will need to be addressed for new products to move into humans. Golding offered several examples of FDA scientific advances that were shared with MCM developers, such as the development of a high-throughput vaccinia virus neutralization assay, which the agency published and shared with multiple manufacturers of vaccinia immunoglobulins and new vaccines. She also highlighted several ongoing areas of research, such as faster or alternative approaches to assessing vaccine potency and new methodologies to evaluate tumorigenic cells to be used as cell substrates in lieu of egg-based vaccine processes (e.g., for tumorigenicity, oncogenicity, unknown adventitious agents). Basil Golding, director of the Division of Hematology at CBER, added that an advantage of FDA research and development is that the agency scientists are then very familiar with the parameters and pitfalls of various assays, and can provide valuable technical assistance to sponsors. David Frucht, chief of the Laboratory of Cell Biology at CDER, also agreed, noting that promoting research (both regulatory and product related) at FDA not only helps overcome development hurdles, but it establishes the subject matter experts at FDA that can provide rational and expedited reviews prior to and during emergencies.

With regard to vaccine development, Frucht emphasized that vaccine potency is a critical product quality characteristic, and the potency assay used should reflect the presumed mechanism of action of the product in humans. This is especially relevant, he said, when the Animal Rule is being used to establish clinical efficacy. He advocated for continued research into the most biologically relevant bioassays and expansion of existing programs. Frucht also said that more in-depth knowledge of product quality early in the development cycle could help accelerate availability of MCMs for potential use under an EUA. The manufacturer should have an in-depth understanding of the critical process parameters and the manufacturing process early on in development. If done late in the development cycle, changes to manufacturing (e.g., scale up) and changes in the producer cell type can affect product quality, resulting in delays.

Alan Shaw, chief scientific officer for Vaxinnate, reinforced Magill's point that one of the major problems of viral vaccine development is the protracted time frame. To help speed development, Vaxinnate is using a platform approach in which the company is inserting antigens onto flagellin (a toll-like receptor agonist). This, he explained, renders highly visible a protein or an antigen that is otherwise basically invisible to the immune system. This technique can be used for a variety of targets. Vaxinnate's primary focus is influenza, but Shaw noted that they are also applying this approach to flaviviruses (e.g., West Nile virus, Japanese encephalitis, yellow fever, hepatitis C virus). There are other relevant platforms that could be applied, and Shaw suggested a platform consortium be formed. He also suggested a viral structure consortium that would develop common databases of structural information characterizing the different classes of virus that are likely to emerge as human pathogens.

Ed Nuzum of NIAID noted that there are numerous challenges for MCM sponsors beyond the Animal Rule: process development, manufacturing, product characterization, and potency assays are just examples. Despite FDA's willingness to communicate early and often with sponsors, in practice it often does not happen for various reasons. He expressed concern that FDA's action teams, discussed earlier by Luciana Borio of FDA, address higher-level, crosscutting, cross-center issues but not the product-specific concerns that sponsors may have. Nuzum supported the idea of working groups or "product acceleration teams" to enhance communication between sponsors and FDA, especially for companies that lack adequate expertise.

Key Messages: Vaccines and Adjuvants

- Manufacturing and other process changes should be reviewed to determine what process requirements are necessary, and creative solutions such as MBFs should be investigated.
- Vaccine potency assays should be further studied to ensure they are biologically relevant and reflect the presumed mechanism of action in humans.

continued

Key Messages Continued

- There is a need for a defined regulatory pathway or qualification process for vaccine adjuvants; broad licensure for adjuvants that can be used with virtually any MCM vaccine should be considered.
- An adjuvant advisory group, including FDA, industry, and academia, could be formed to convene workshops on adjuvant-specific issues such as safety biomarkers and adjuvant combinations.
- Consortia groups should be developed to address issues such as:
  - platform technology, and
  - viral structure, to develop common databases of structural information characterizing virus classes likely to emerge as pathogens.

SYNTHETIC AND COMPUTATIONAL BIOLOGY AND PLATFORM TECHNOLOGIES

Synthetic Biology

John Glass, senior scientist at the J. Craig Venter Institute, defined the emerging field of synthetic biology as the production of biological life, or essential components of living systems, by synthesis, to make new organisms with extraordinary properties. What makes synthetic biology possible is the fact that genes can be synthesized using the four bases that make up DNA. This capacity to synthesize DNA is changing the field of biological research, he said.

Glass described six examples of how synthetic biology can be used to advance MCM development. First, synthetic bacterial cells can be used to produce live attenuated vaccines (or the equivalent). The first bacterial cell with a chemically synthesized genome was *Mycoplasma mycoides*. This was the result of two technologies that were developed at the Venter Institute: the "Gibson Assembly Method" of rapidly and efficiently assembling oligonucleotides by overlapping synthetic DNA molecules into larger molecules, and genome transplantation, the capacity to transfer a whole, naked bacterial genome from one organism into another so the recipient cell is converted into the same organism as the donor cell (Figure 3-4).

With this approach, for example, one could synthesize pathogenic bacteria devoid of all virulence factors, which could then potentially be used as live attenuated vaccines, Glass said. The transplanted genomes could contain only those genes necessary to generate an immune response and keep the cell alive. Adjuvants could be built into the organisms, and if desired, suicide genes could be inserted so that the organism could only survive for a specified number of generations. Another example would be the synthesis of gram-negative bacteria that produce outer membrane

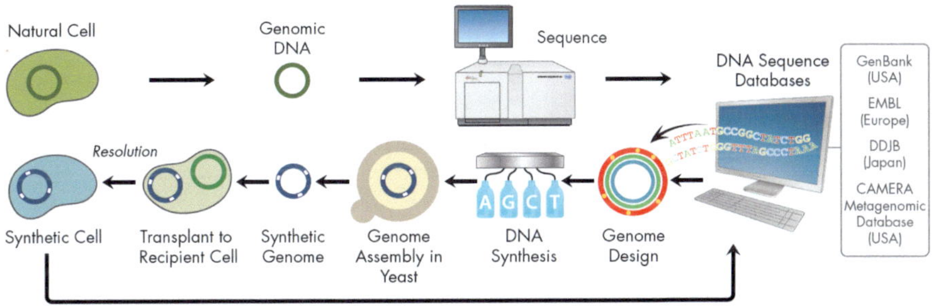

FIGURE 3-4 Synthesis of bacterial cells.
SOURCE: John Glass. 2011. Presentation at IOM workshop; Advancing Regulatory Science for Medical Countermeasure Development.

vesicles containing immunogenic proteins from a variety of pathogenic bacteria, kind of an "omnibus vaccine," Glass said. This allows for immunization with membrane-bound proteins without exposing people to live cells.

Second, synthetic biology offers the potential to make influenza virus vaccine production better and faster. Using reverse genetics to synthesize hemagglutinin and neuraminidase can decrease the time needed to achieve a pure culture of vaccine seed stock from 35 days to 7 days (Figure 3-5). However, Glass said, while the technology to produce influenza virus vaccines has advanced greatly over the last 20 years, the regulatory processes for licensing new pandemic and seasonal influenza virus vaccine have not progressed in concert.

Other examples of potential uses of synthetic biology for MCM development are:

- The synthesis of a bacteriophage for use as an antibacterial therapeutic.
- Microbial manufacturing of drugs currently obtained from scarce natural sources (e.g., the antimalarial drug, artemisinin, originally extracted from Chinese wormwood bark, is now synthetically engineered).
- Synthesizing antigens for use in diagnostic assays and evaluation of therapeutics and vaccines (e.g., expressing protein epitopes from four different *Borrelia burgdorfi* genes on one recombinant protein to make an effective *B. burgdorfi* diagnostic).
- Finally, synthetic biology can be used to develop new tools and approaches to discover novel therapeutics.

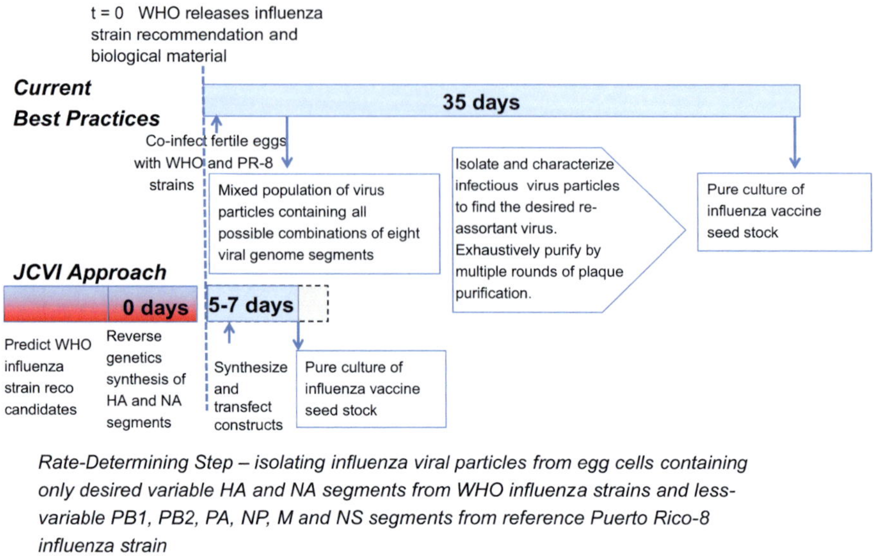

FIGURE 3-5 Accelerating seasonal influenza virus vaccine development.
NOTES: HA, hemagglutinin; JCVI, J. Craig Venter Institute; NA, neuraminidase; WHO, World Health Organization.
SOURCE: John Glass. 2011. Presentation at IOM workshop; Advancing Regulatory Science for Medical Countermeasure Development.

In closing, Glass said that fast, accurate DNA synthesis of genes and genomes can eliminate the time and effort needed to obtain a natural DNA template for known pathogens or unknown or newly described infectious agents. Accurate, inexpensive synthetic DNA makes it possible to create many new therapeutic and diagnostic tools that were previously impossible or impractical to build. Engineering of microbes with extraordinary properties will soon be done in days or weeks.

However, Glass predicted, this accelerated pace of new biotechnology development may swamp FDA capacity to evaluate what he said would be "an avalanche of new drugs, vaccines, therapeutic, and diagnostic assays." To help advance development, Glass suggested that older licensed compounds, vaccines, and assays that are made using new synthetic biology tools should be considered equivalent to their predecessors.

Computational and Systems Biology

Modern day virologists and immunologists have access to substantial technology, information, and computational infrastructure, said Michael Katze, Professor of Microbiology at the University of Washington, noting, however, that even though the tools are there, scientists are ill equipped and ill trained to deal with the data. In addition, it is still not fully understood how a virus kills a cell, or how host attributes contribute to response. For example, when using model systems, if key findings do not align, is that a problem with the model or is it related to other attributes of the host?

Handling next generation sequencing, microarray, proteomics, and other "-omics" data requires development of a sophisticated information technology infrastructure, which, Katze pointed out, is expensive and complex. These new tools and technologies are important to virologists because they aid the study of the global impact of virus infection on host gene expression; and they can be used to discover cellular regulatory pathways targeted by viruses, identify new cellular targets for antiviral therapies, and develop new vaccines. Katze proposed an integrated approach to infectious disease, combining traditional histopathological, virological, and biochemical approaches with functional genomics, proteomics, and computational biology.

As an example, Katze described a systems biology approach to studying the 2009 H1N1 pandemic across species. He noted that H1N1 does not generally kill swine, and perhaps understanding why pigs do not die will provide a better understanding about host defense against influenza. Studies of the transcription of immune-related genes across mice, macaques, and pigs suggested that the nature of the immune response in each species may be quite different. Overall functional analysis showed significant alteration in immune response genes in all three species. Although the numbers of genes changing was similar, the precise genes changing were very different.

Katze suggested that applying systems biology to vaccine development could potentially help detect early signatures of efficacy and offer information about why a vaccine fails. Once host factors of virulence and pathogenesis are identified, systems biology could be applied to drug development, for example, to "score" drugs based on response to treatment, or to better understand mechanisms of action (including off-target effects of drug treatment and toxicity) through gene expression studies.

Computational biology will be key to modeling and predicting host response, Katze said, adding, however, that several challenges exist. As the landscape for certain high-throughput technologies is still being defined, there is a need to be able to accommodate constant evolution. Cost is an issue, as is communication; in other words, there is a need to connect the right people (computational scientists, mathematicians, biologists) as well as teach a new generation of scientists a new vocabulary and a new way of viewing science. To begin to address these challenges, Katze recommended:

- Early integration of high-throughput data collection in drug and vaccine development as a mechanism to understanding global impact, off-target effects, and biomarkers for efficacy.
- *In silico* screening for drug-drug interactions and as a tool for novel drug discovery.
- Increased interdisciplinary crosstalk between computational scientists and bench scientists to define standards for study designs.

In discussion, a participant raised the issue of training the next generation of the workforce to advance regulatory science outside the context of a particular product. Katze noted that universities have started offering interdisciplinary programs in computational biology where previously there was very little interaction between computer science and biology. A participant from FDA added that the agency has been putting resources into science computing capacity and is training agency reviewers and researchers to be able to use them.

Platform Technology

As an example of the use of platform technology to advance MCM development, Patrick Iversen, senior vice president of research and innovation at AVI BioPharma, described his company's approach for the rapid development RNA-based therapeutics. AVI's platform is based on the development of translation-suppressing oligomers that target single-stranded RNA (which could be from a host cell or from the pathogen), preventing the assembly of the ribosomal complex on the mRNA transcript, thereby preventing the production of a specific protein.

AVI has developed a predictable way of designing the oligomers, which makes the platform very flexible and allows for very rapid response. They have defined both the optimal position in the transcript, and the optimal length of the oligomer, and are developing a database of oligomers for a growing list of viral and bacterial targets and host genes. This knowledge base, Iversen predicted, would allow AVI to develop a putative solution to a new threat in a matter of hours.

Iversen noted that AVI currently has open INDs for oligomers for Ebola and Marburg viruses. Studies in mice, guinea pigs, and nonhuman primates have shown significant protection (i.e., survival). Crossover studies confirmed the specificity of the oligomers for the intended target (i.e., the Ebola virus oligomer was not effective against Marburg virus, and vice versa). Other endpoints investigated included dose-dependent survival increases, reduction in clinical signs, reduction in viremia, increase in platelet count, and improvement in both hepatic and renal markers of toxicity.

In closing, Iverson raised several questions regarding animal studies and human safety testing. For animal models, he asked, how should a viral challenge strain be chosen? For example, would it be better to use Marburg Angola or Marburg Musoke? Quasi-species characterization could reveal that there are elements or portions of both viruses in every outbreak. And the next outbreak will be a new quasispecies. "Deep sequencing" technology, he suggested, could provide insight into how to choose challenge strains.

Iversen also questioned whether the use of healthy volunteers for safety assessment is necessary for MCM development. He noted that in normal healthy volunteers, the dose limiting toxicity may fall below the anticipated therapeutic dose. How should that limitation be interpreted; what distance between anticipated therapeutic benefit and dose limiting toxicity will be tolerable? Also, how should the size of the required human safety database be calculated? He asked, if these MCMs will never be used unless there is an outbreak, and will be used only used under an EUA, is a human safety database needed?

Discussion

William Fogler, senior director of portfolio planning and analysis at Intrexon Corporation, pointed out that the need for rapid response generally occurs under worst-case scenarios, often in association with compromised infrastructure. While these synthetic, computational, and platform technologies offer tremendous promise to respond rapidly to a pathogenic threat, they must be scalable and deliverable under such a scenario. He suggested that there are additional technologies that exist in terms of generating DNA vaccines, in which modular components can be predesigned, stored, and ready to assemble on short notice. Other modules could be devised in which immune-enhancing agents could be quickly assembled. These modules in the structure of a DNA vaccine can be under the control of inducible promoters, so that following injection of the vaccine, an activating ligand (e.g., a small molecule) would be taken orally to "turn the vaccine on," and upon removal of the ligand, it would be "turned off." This also offers the possibility of a needle-free vaccine-boosting mechanism, Fogler said.

Mendrick said that researchers at FDA are looking at these emerging technologies and are trying to anticipate and solve some of the questions that may arise. For example, NCTR has a nanotoxicology core facility that is looking at genetic toxicity assays to evaluate the carcinogenicity of nanoparticles.

Harvey Rubin, executive director of the Institute for Strategic Threat Analysis and Response at the University of Pennsylvania, emphasized that computational biology is not simple mathematics. The scale of computational biology spans angstroms to kilometers, and nanoseconds to millennia, he said. The processes are very complicated, and include, for example, deterministic, stochastic, continuous, discrete, or hybrid processes. With regard to organization, the system could be structured, unstructured, or homogeneous. There are complexities and interdependencies that make modeling biological systems especially difficult, Rubin said. Motivations to do complicated mathematical modeling include the need to predict something (e.g., protein structure, epidemiologic patterns), to design something (e.g., new molecular structures, new controllers and regulators, new phenotypes), or to interpret something (e.g., data, patterns).

Rubin highlighted several research priorities that can help populate some of these mathematical models:

- There are many model-specific questions that need to be answered, such as what are the effects of interventions on infectivity, and what are the effects of disease and interventions on immunocompromised hosts?
- There is also general research needed on organizational structures, risk communication strategies, interdependencies (e.g., how the environment, economics, or politics impact the model), and health impact information.
- Also to be resolved is who should be funding this work—NIH, the National Science Foundation, DARPA, FDA, or industry.

DIAGNOSTICS

Significant resources are dedicated to identifying and characterizing an emerging biological threat, said Daniel Wattendorf, program manager in the Defense Sciences Office of DARPA, but rarely is there subsequent broad distribution of new diagnostic assays for the identified threat to point-of-care settings. In cases where the decision to quarantine or treat is time sensitive, the turnaround time to ship samples to a reference laboratory is prohibitive.

Wattendorf cited several barriers to more rapidly fielding diagnostics for emerging threats. In some cases, the diagnostics platforms have not been made suitable for use in distributed settings. As an example, Wattendorf pointed out that PCR has been in use since 1983, yet no PCR-based diagnostic test is approved for a physician office setting. Additionally, if diagnostic tests are not already in place before an emergency, it is very difficult to get physicians to employ them in a crisis if they do not have prior experience with the test or have not been shown evidence of utility. In the absence of specific diagnostic tests for emerging threats, there is interest in developing panels of early detection biomarkers that could detect a host immune response before an individual begins to exhibit symptoms of a disease.

Sample collection is another challenge for diagnostic testing. Current biospecimen collection generally involves collection of wet samples, such as through test tubes, which requires that the patient have access to medical personnel (e.g., a phlebotomist) who can collect the sample, and which also may require cold storage. There is also the option of taking dried blood spots on filter paper, but according to Wattendorf such samples have limited use. In this regard, Wattendorf suggested that a role for regulatory science would be the development of new formats for simple, self-collected biospecimens, formats that would be optimized for specimen source (blood, urine, etc.) and analyte class (specific proteins, types of RNA, etc.), and would be stable during storage to facilitate functional assays.

Wattendorf also noted that currently, teams of experts travel to a site, collect samples, and return to CDC or the DoD to run tests and identify the new threat. He suggested that, instead of moving the sample, it could be possible to move the data electronically. The use of highly multiplexed platforms could facilitate local testing, and the data could then be sent to a central facility for analysis. This would be faster and would provide distributed diagnostic capability where there is unmet need.

In summary, Wattendorf listed several questions for discussion:

- Can universal sample storage formats be developed for dried or near-dried self-collected biospecimens that show equivalence to fresh samples?
- Can highly multiplexed protein or molecular diagnostic platforms be developed that are suitable for use in a physician office base setting, from which data could then be sent for interpretation by highly trained laboratorians at a remote site?
- Are measurements of immune or metabolic status useful in the absence of a diagnostic test for a specific pathogen? If so, what should be measured? Could it be measured at the point of care? And, as it is not specific to a given disease, what would be the regulatory pathway?

In the panel discussion, Charles Daitch, CEO of Akonni Biosystems, said that from a technical perspective, the capability to communicate from remote sites to a central facility already exist, and it would be straightforward to develop and implement ways to communicate using either raw or processed data. Sally Hojvat of CDRH concurred and suggested that this would be covered under existing regulations that address electronic records and the transfer of data from an instrument at a clinical site to a central facility for analysis (21 CFR 11). She cautioned that it would be necessary to demonstrate the accuracy and traceability of the results of a test performed remotely by an unqualified individual.

Panel moderator Bruce Burlington, an independent consultant, questioned how it could be determined that an immune status test was relevant for many different illnesses. Would test developers need to undertake a variety of disease challenges? Hojvat responded that it could be considered more of a prognostic type of marker, and such data would be one way FDA could begin to assess the test. With regard to its commercial value, Daitch said that the market for such a test is not obvious. A test that predicts, based on immune status, that someone is in the early stages of an infectious disease might be useful, for example, for astronauts about to go on the space shuttle or for troops about to be deployed, he said. Burlington added that it could also be used in an epidemic for health care workers or other first responders.

Participants discussed the potential for commercial assays on multiplex platforms to be used as epidemiological surveillance tools (as opposed to diagnostic tests where results are reported back to the patient). Hojvat suggested that companies could aid the surveillance effort by developing cassettes for biothreats for their multiplex systems. Daitch and David Ecker, founder of Ibis Biosciences, agreed it would be possible, but noted that key challenges would be validation of the test for broad groups of organisms and ensuring that data could be transferred over a secure network to somebody who has the capability to interpret the data correctly.

Participants also discussed the concept of an evolving label. Performance characteristics of a diagnostic test need to be defined in terms of sensitivity and specificity, but a challenge is how to present that information in the label when the background prevalence of what is being tested for is almost zero. It would be helpful if, as the threat emerges, new information and data based on use could be made available rapidly. Hojvat responded that FDA has the technology to do that, and there is an ongoing electronic labeling project.

In summary discussion, participants observed that it is important to remember that diagnostics are also MCMs. Several options for more efficient use of diagnostics were suggested, including the development of new formats for collection, transport, and stable storage of biospecimens, and the development of highly multiplexed testing platforms for local site use, with data then sent electronically to experts at a central facility for analysis. It was also noted that rapid diagnostics could improve the efficiency of antimicrobial trials, allowing for enrichment of the population with patients infected with resistant organisms.

STATISTICAL TECHNIQUES

The goal of clinical trial simulation in drug development programs is to reach a decision faster, cheaper, and with greater certainty, explained Stephen Ruberg, distinguished research fellow and scientific leader in advanced analytics at Eli Lilly and Company. Companies seek to "kill" ineffective or unsafe investigational drugs sooner and advance potentially effective drugs as quickly as possible and at the lowest cost possible. Clinical trial simulation allows for examination of a broad range of clinical trial designs, decision rules, and analysis methods. In simulations, models can be used to create virtual patients that are then randomly selected for inclusion into *in silico* clinical trials using sophisticated software tools. These models for virtual patients can be PK/PD models, empirical statistical models of response over doses and time, or mechanistic disease models. Known design and analysis parameters can be controlled (e.g., sample size, number of doses or visits, analysis strategies for testing hypotheses or estimating key drug effect parameters), and a range of possibilities for unknown parameters and factors that cannot be controlled can be assessed (e.g., drug effect, true dose response curve, adverse event rate, placebo response, dropout rate). Dozens of combinations of factors are typically evaluated with the goal of selecting the design and analysis parameters that will minimize false positive and false negative findings in the drug development program.

From a regulatory science perspective, Ruberg said, this will require training of FDA staff on the use of simulation tools, some of which are becoming commercially available. FDA statistical and medical reviewers will have to understand and accept modeling simulation as a tool for study design. Simulation trial designs generated may not look like classic trial designs or may not have theoretically or mathematically described properties, he said. This is of particular concern when designing phase III trials due to the need to control the type I error (false positive) rate at 0.05. As this cannot always be done analytically, Ruberg asked whether FDA will accept simulated results in lieu of analytical proof. He noted that the FDA draft guidance on adaptive designs[3] is a substantial step forward in helping the industry understand how best to move forward with innovative trial designs. Another topic for consideration is the simulation of the

---

[3] See *Guidance for Industry Adaptive Design Clinical Trials for Drugs and Biologics* (Draft Guidance) http://www.fda.gov/downloads/Drugs/GuidanceComplianceRegulatoryInformation/Guidances/ucm201790.pdf (accessed June 9, 2011).

sequence of clinical trials spanning an entire clinical drug development program, which, Ruberg said, companies could realistically be doing in the next couple of years.

A goal in drug development is to use as much data as possible—current or historical—to make decisions on drug safety and efficacy. Current practice in the vast majority of phase III clinical programs is for each clinical trial to stand on its own as an independent piece of evidence in the evaluation of a drug's effect. This is a *frequentist* statistical approach. Eli Lilly, Ruberg said, is currently implementing *Bayesian* methods for phase I and phase II trial design and analysis. There are many ways in which Bayesian methods can be used in clinical drug development. One example presented by Ruberg is a Bayesian augmented control design, in which control group data from the current prospective study is supplemented with historical control data. This allows for smaller trials (saving both time and resources) and for more enrolled patients to be allocated to treatment groups.

While the use of Bayesian statistical methods is a technical topic, Ruberg opined that the largest barriers to implementation are social. There will need to be changes in philosophy and mind-set within some FDA centers and other regulatory agencies around the world. There are also legitimate scientific debates relative to the choice of historical data to include in analyses and how to weigh those data relative to data generated from a new study, he added. From a regulatory science perspective, Ruberg said that the use of Bayesian approaches for phase III confirmatory trials would require in-depth sponsor-agency discussions at the end-of-phase-II meeting or sooner.

Important to the use of Bayesian approaches is the development of a comprehensive data element dictionary. Data element standards allow for more efficient collection of data and routine use of standardized software. More importantly, common data element standards allow for the simple, rapid integration of data from multiple sources, facilitating more comprehensive statistical analysis in order to draw the best scientific conclusions possible. Such a dictionary should be maintained by a central authoritative group, Ruberg said, and must be free, broadly accessible in electronic form, and downloadable for use within IT systems. He acknowledged the various ongoing standardization efforts (e.g., CDISC, HL7),[4] but said that

[4] The mission of the global, nonprofit, multidisciplinary Clinical Data Interchange Standards Consortium (CDISC) is to "develop and support global, platform-independent data standards that enable information system interoperability to improve medical research and related areas of health care." See http://www.cdisc.org/. Health Level Seven International (HL7) is a nonprofit "standards-developing organization dedicated to providing a comprehensive framework and related standards for the exchange, integration, sharing, and retrieval of electronic health information." See http://www.hl7.org/ (accessed June 9, 2011).

data element standards needs to go deeper in terms of specificity, broader in terms of accommodation of all therapeutic areas and measurements, and faster in terms of development and deployment.

In closing, Ruberg offered several ideas to advance the use of trial simulation and Bayesian statistics, and the standardization of data elements:

- For study design, Ruberg suggested adaptive/pooled studies as a way to more rapidly and uniformly test compounds. Such studies use a single trial design with a common control group that allows companies to insert their drug into a perpetually ongoing trial (such as the I-SPY 2 breast cancer clinical trial).
- Ruberg also directed participants to the Drug Information Association (DIA) Working Group on Bayesian Methods. Ruberg noted that Bayesian approaches were discussed in a recent National Research Council (NRC) report on how to handle missing data in clinical trials (NRC, 2010), and he suggested that the National Academies conduct a study to evaluate the use of Bayesian methods in clinical trials, with particular emphasis on phase III confirmatory trials.

In panel discussion, there was much discussion about the use of Bayesian statistical methods for analysis of clinical trials. Jeffrey Wetherington of GSK said that his company has made significant use of Bayesian methods for phase II proof-of-concept studies and dose-ranging studies, and he estimated that use of these methods has saved the company nearly $15 million on study costs over the past year. Similar to Eli Lilly, Wetherington said, GSK uses augmented control groups, decreasing study sample sizes by several hundred people. Bayesian methods provide very interpretable results, he said.

Estelle Russek-Cohen, acting director of the Division of Biostatistics at CBER, noted that CDRH frequently uses Bayesian analysis in the context of device modification submissions. She added that CBER has seen submissions with Bayesian and adaptive designs, primarily in phase I and II studies, many of which have been oncology studies. For phase II studies, a variety of skill sets are needed when considering the benefits and risks of the analysis approaches (e.g., medical officers, statisticians). Russek-Cohen said a concern with historical controls is how far back to go if the standard of care is changing. A question for consideration is whether, in the context of MCMs, there is a real and compelling need for these approaches.

Panel moderator Burlington noted that the toxicology community routinely uses historical controls, pooling data from control animals from many experiments. Russek-Cohen responded that pooling of historical control data is used for safety assessment as there is often not enough power in individual studies, but it has not yet been done for efficacy studies, in part because FDA statutes call for adequate and well-controlled trials. In a phase II environment, it makes sense for industry to find ways to pool control information across companies pursuing similar projects.

Wetherington said that drugs such as anti-infectives can have small niche markets and limited profit margins. Using Bayesian-type designs for phase III studies, especially when comparing the novel agent to a well-characterized standard of care, could save time and money, and get products to patients more quickly.

Participants discussed the potential for use of simulation and Bayesian approaches as the basis for approval of an MCM in anticipation of future use. Goodman responded that part of the FDA MCM initiative is to consider novel approaches, and the agency is open to these possibilities. He encouraged developers of MCMs to discuss this with their FDA review team as part of their product development planning.

Burlington questioned whether FDA could mandate or incentivize companies to submit their data in conformation with data element standards. Russek-Cohen responded that implementing standards is part of the broader FDA initiative. Ruberg suggested that the National Library of Medicine or FDA could take the lead on pushing forward with data element standards. A participant noted that CDISC is approaching standards development disease by disease. In response, Ruberg suggested that there could be a working group of experts in MCMs to define what generally needs to be measured and start discussing standard data elements.

In summary discussion, the statistics of diagnostics and dealing with false positives was also considered. A participant said that CDRH has asked developers of new multiplex diagnostic assays to offer ideas about how to handle false positives, for example if three or four positives were found where one was expected. A participant said to keep in mind the primary question the assay is answering: Is it diagnosing an individual or determining if there is an outbreak? It was noted that there is an interagency meeting being planned on this issue.

Several workshop presenters and discussants noted that Bayesian statistical methodology can be used for both study design (e.g., supplementing the control group data with historical control data) and analysis (of both actual and simulated trials). Workshop participants offered suggestions for themes and future directions with respect to statistical methodologies and data analysis. The following individual suggestions were made:

- Training in Bayesian approaches and causal mechanisms of actions will be needed for both scientists and the public.

- The use of Bayesian approaches would be enhanced by the development of common data element standards (e.g., to facilitate pooling of data across studies).
- Christian Macedonia, medical sciences advisor to Admiral Mullen, the chair of the Joint Chiefs of Staff, raised the idea of electronically tagging every piece of information obtained in biomedical research (e.g., date, time, group, unique animal identification, institution) so data from large multicenter trials could be traced back, even years later, for further analysis. He likened this to the way electronic data are broken into packets and tagged for transfer across computer networks, to be reassembled at the other end.
- There was also interest in platform approaches to health data software design, for which many applications or "apps" could be developed. These could be used for collection, management, and analysis of electronic health data, specifically for monitoring of adverse events.

4

MCM Regulatory Science Needs for At-Risk Populations

CHILDREN

David Siegel of the National Institute of Child Health and Human Development (NICHD) reviewed a list of pediatric-specific vulnerabilities relevant to exposure to CBRN agents and development of MCMs (Box 4-1). He expressed concern that children and pregnant women are often labeled as "special populations." Children, Siegel countered, represent a very large segment of the population. They do not have "special needs," even though they may be vulnerable. In addition, 10 percent of all women of child-bearing age are pregnant at any given time. This stereotype has resulted in research focused on adults, with studies on the "special populations" set aside for a later time, if funding permits. There has been a lack of funding at all levels of the MCM development process regarding children and pregnant women, Siegel remarked.

Another issue Siegel raised is off-label use of treatments. Regulatory issues have created special challenges for the deployment of appropriate biodefense medicines for children. In regular pediatric medical care, 50 to 75 percent of pediatric medications are used off-label. This lack of approved pediatric labeling for certain indications significantly affects federal jurisdiction and deployment of MCMs in the SNS, Siegel pointed out. He posed several questions for consideration: Should pediatric labeling be required for deployment of therapeutic agents to the SNS? Should Congress amend the EUA statute such that key MCMs lacking the requisite use labeling can be forward deployed?

Adding to the challenge is the widespread thinking that pediatric

BOX 4-1
Pediatric Vulnerabilities

- Children are known to be at greater risk following exposure to CBRN agents.
- As children are lower to the ground, they are exposed to an increased concentration of CBRN agents.
- With higher respiratory rates and lesser volumes than adults, a child will inhale a greater dose of agent.
- Children have smaller diameter airways, anatomic subglottic narrowing, omega-shaped epiglottic structure, relatively large tongue size, and less rigid ribs and trachea which make them more vulnerable to agent-induced pathology such as bronchospasm, copious secretions, and pulmonary edema.
- A child's smaller mass alone reduces the dose of chemical agent required for toxic or lethal effects.
- Nerve agents penetrate the blood brain barrier more easily in children than adults, and children may only exhibit central nervous system (CNS) effects.
- Animal studies have shown that the lethal dose of nerve agent in an immature animal versus an adult animal is 10 percent.
- Young children, especially less than 4 years of age, are more prone to develop seizure disorders secondary to hypoxia or other CNS insult.

SOURCE: Siegel presentation.

studies should not be done until all of the adult studies have been completed. In some cases, this is appropriate, but in other cases it is not, Siegel said. If animal studies are completed, it is not necessary to wait until phase III clinical trials are completed before starting juvenile animal studies. There are challenges to performing pediatric and obstetrical drug trials due to realistic (and nonrealistic) patient safety issues. In some circumstances, the pharmaceutical industry has been reluctant to produce therapeutic agents or devices for children because of a lack of profitability. This is being addressed to some extent, Siegel noted. There is also a shortage of researchers at all levels who are capable of performing studies with children or pregnant women.

Moving forward, Siegel supported efforts to establish an obstetrics/pediatrics section or working group at BARDA to examine the current contents of the SNS, be aware of up and coming MCMs, be part of the MCM prioritization process, and make obstetric/pediatric study recommendations for the necessary PK, efficacy, and safety data. FDA, he said, should be actively involved in this section, working proactively with academia and industry sponsors to expedite MCM development. Increased harmonization and sharing of data by FDA, industry, and academia is also needed. Siegel called for expansion of the Best Pharmaceuticals Act for Children to cover MCM development for children. Other recommendations offered by Seigel included the following:

- Prioritize funding of a systems approach for pediatric formulation development.
- Encourage timely development of appropriate juvenile or pregnant animal models (and determine whether juvenile animals are really needed for a given specific indication).
- Increase development of pediatric virtual and organic modeling capabilities.
- Encourage development of pediatric biomarkers.
- Utilize the forthcoming central institutional review board (IRB) for disaster-related studies as a platform for prospective studies of children.
- Work closely with FDA to ensure that these studies are designed to collect the requisite information.

PREGNANT WOMEN

Pregnancy is the most dynamic period of human growth and genomic expression, making it a period of some vulnerability, said Christian Macedonia, medical sciences advisor to Admiral Mullen, the chair of the Joint Chiefs of Staff. Genomic expression, he noted, is not just about making proteins, and during pregnancy all aspects of the genome are in play. Pregnancy is also a period of immune modulation, including immune tolerance and immune suppression. However, immune suppression during pregnancy is not on par with, for example, that resulting from chemotherapy. While pregnancy is often referred to as a "delicate" condition, Macedonia emphasized that pregnant women are extremely durable. Pharmacokinetics in pregnancy is uniquely challenging. Increased volume of distribution, altered protein binding, increased glomerular filtration rate, and other maternal changes complicate dosing determinations.

From a more societal perspective, Macedonia pointed out that pharmaceuticals and vaccines are generally not tested in pregnant human patients. Drug safety in pregnancy is typically obtained through post-market surveillance.

With regard to benefit-risk assessments, Macedonia noted that conventional wisdom has been that humans are rational creatures and make decisions based on what is the best outcome. However, studies in game theory and behavioral economics suggest that humans are not entirely rational and make decisions based on a calculus of what is not just the best outcome, but what is the most likely outcome. He described the "regulator's dilemma" (a la the prisoner's dilemma from game theory) as this: From which perspective does the regulator derive his or her assessment? Who are they regulating for (e.g., the physician, the maternal patient, the fetal patient, society, the manufacturer, the agency, themselves)? Regulatory decisions can have different risks and consequences for each of these stakeholders.

Macedonia offered several suggestions to help meet the MCM regulatory science needs of pregnant women. Open, transparent, and interactive public education on the benefits and risks of MCMs is needed. He called for leveraging the power of the social network, both as a means for education (e.g., though interactive apps) and for understanding the perspectives of the public. There is also a need for greater education of scientists and policy experts on the value of exploratory data methods (i.e., there are other valid options besides a randomized controlled trial). Finally, he recommended greater investment in point-of-use diagnostics to better define the at-risk population and the right dose of medication or vaccination needed.

In discussion, Pravin Jadhav of FDA's CDER explained that FDA has a pediatric decision tree to guide decisions regarding what kind of data, and how much, is needed to approve a drug for pediatric use. There would be no difference, he said, between how the agency would handle drugs for pediatric MCM use compared to any other pediatric drug (although he acknowledged that MCM development may rely more on data from animals, as well as PK and systems biology data). As an example, he said that in 2009, the agency was writing the EUA for Peramivir for H1N1 influenza, and there were no data available on children from birth to 18 years of age. To address this, FDA reviewed data from other similar drugs that, like Peramivir, were eliminated through a renal route. Those data, along with what is known about the developmental biology of the kidney, allowed the agency to derive pediatric dosing recommendations. Jadhav noted that the EUA clearly states that the dosing was derived based on modeling and simulation. In 2010, data from children treated in Japan indicated that the prediction was quite accurate; in 2- to 9-year olds the dosing was off by 15 percent, and in ages 9 to 18 years it was on target.

Nancy Messonnier of CDC described the use of anthrax vaccines in children as a case study. Anthrax vaccine was licensed in a general use protocol for preevent vaccination. Following a bioterrorism event, it would be used in accordance with an EUA; however, as there are no data in children, the EUA would be limited to adults. The CDC Advisory Committee on Immunization Practices and the American Academy of Pediatrics Red Book Committee state that in the event of an anthrax event with inhalational exposure, a benefit-risk analysis should be made, and children should be given anthrax vaccine if warranted. CDC and FDA are developing a strategic framework for how children would be vaccinated in an event, including how immunogenicity and safety data could be collected.

Messonnier expanded on the communication of benefits and risk raised by Macedonia, noting that in a public health emergency there is a need to communicate to large populations. There may not be opportunities to have one-on-one conversations with patients. When communicating to parents about the risks and benefits of vaccinating their children, how can we be clear and unambiguous, but also appropriately identify that there is a lack of data?

Lisa Mathis of CDER noted that due to the Best Pharmaceuticals for Children Act and the Pediatric Research Equity Act, among other developments, nearly 50 percent of products are now labeled for pediatric use, up from 25 percent. Unfortunately, she noted, there has not been that same level of advocacy in support of pregnant and lactating women. Mathis said that FDA's proposed pregnancy and lactation labeling rule (currently being finalized) would remove the five letter categories from labeling and would provide more clear information about what is known about use in pregnancy, from both human and animal data.[1]

Robert M. Nelson of FDA said that the assumption that children are a vulnerable population, not only in the physiologic sense but also in the ethical sense, has resulted in additional protections for children who are enrolled in research. Unfortunately, he said, this has often resulted in a protection *from* research, rather than protecting children *through* research. The goal of research should be concurrent licensure for both adults and children so that children are not placed in the vulnerable position of receiving drugs off-label.

There are two general pathways in the additional protections for pediatric research. If there is not a prospect of direct benefit to the study subject, the risk of the intervention must be minimal. For higher-risk products under investigation, there must be a prospect of direct benefit to the subject. For example, much of the Pralidoxime data was generated through pesticide exposure (specifically, clusters in New York City where children ate rat poison imported from abroad, which was composed of organophosphates, not Coumadin). Data were generated by clinicians who were administering Pralidoxime off-label to these children. The decision to administer was based on clinical grounds, and the research (basically a blood test) was secondary, and a low risk to the patients.

During a public health event, there are many operational challenges to simultaneously conducting research: Where will the work be done?

---

[1] See http://www.fda.gov/Drugs/DevelopmentApprovalProcess/DevelopmentResources/Labeling/ucm093307.htm (accessed June 9, 2011).

How do you pre-position assets around the IRB approval? These operational issues are solvable, Nelson said. Research in the preevent setting, where children are not at risk, is more challenging. There are some situations where neither pathway applies. In those cases, it was proposed that for studies that were scientifically sound and ethically appropriate, there could be a level of federal review. However, the decision whether to conduct the trial is a separate one from the parental decision whether to enroll their children in such a trial if there is no direct benefit to the child, Nelson noted.

Nelson added that although FDA does not have regulations governing research involving pregnant women, HHS human research regulations (45 CFR 46, Subpart B) provide for additional protections for pregnant women and fetuses. If an intervention is not for the direct benefit of the pregnant woman, then the risk to the fetus must be minimal. Again, according to Nelson, this regulatory framework constitutes a barrier to preevent studies of MCMs.

Participants discussed further the types of studies or data that would be needed for development of MCMs for children or pregnant women. Mathis said that "dose is everything." A good deal of progress has been made for pediatrics because of the Best Pharmaceuticals for Children Act and the Pediatric Research Equity Act, but we need to commit more time and people to studying pediatrics, earlier on in the process. For pregnancy, it would be reasonable to assume that efficacy could be extrapolated, but PK data are most urgently needed. Nelson advocated for an open public deliberative process around both the science and ethics of conducting preevent trials.

5

Crosscutting Themes and Future Directions

ADVANCING REGULATORY SCIENCE[1]

To set the stage for discussion of next steps for regulatory science at FDA, workshop co-chair John Rex, of AstraZeneca, outlined the key crosscutting themes that emerged over the course of the meeting. Throughout the workshop, stakeholders and participants from all sectors emphasized that collaboration is essential for advancing the development and evaluation of MCMs. No single partner has all of the tools. Participants discussed how, in the current research environment, "you get what you reward," and the MCM enterprise needs to find ways to reward flexibility and innovative thinking. Education and training were also highlighted as being essential to advancing MCM development (including scientific training, leadership development, and education of the public). Participants discussed the practical and ethical limitations of conducting clinical trials for MCMs (in both the general population and in at-risk groups such as children[2] and pregnant women), and the available alternative regulatory mechanisms to demonstrate efficacy. In this regard, there was much discussion of the Animal Rule, with a particular focus on the challenges of validation of animal models and establishing true correlates of efficacy.

Rex summed up workshop discussions regarding metrics of success

[1] This subsection is based on the summary remarks provided by workshop co-chair John Rex of AstraZeneca.

[2] Nelson of FDA advised participants that the agency is planning a workshop focusing on the ethical issues of pediatric MCM development for the first quarter of 2012.

in MCM development by noting that definition of metrics is challenging. He commented that there is limited opportunity to assess the true public health benefit of an MCM; success cannot simply be measured by the number of MCM approvals (as not all products will or should succeed). Participants offered a variety of suggestions for metrics, from a goal of adding a defined number of new, approved MCMs to the SNS within a defined time period; to smaller, incremental steps such as developing an assay that solves a key problem, thereby reducing time and/or cost of development; to finalizing MCM-related FDA guidance documents; to approving a product under the Animal Rule. It was repeated throughout the workshop that providing a clear regulatory pathway forward can foster innovation and enhance the quality of sponsor submissions; this, it is hoped, would lead to increased speed of review and success of applications.

From a defense perspective, U.S. troops face threats around the globe, not just traditional biothreats, but endemic diseases as well. It is important to remember that MCM development must include not only vaccines and therapeutics, but also point-of-care diagnostics (for both organism identification and drug resistance profile). An ongoing challenge for both civilian and military populations is getting "the right product, in the right place, at the right time, for the right individuals." In this regard there was discussion of pre-positioning tools (such as diagnostics or mobile manufacturing capability).

Rex added that in the end, for any countermeasures to be effective, the public must accept and use them. In this regard, there is a need to educate the public about advances in regulatory science (e.g., approval and use of products that have not been tested in humans, benefit versus risk during an emergency versus routine medical care). There was interest in leveraging social networks and developing educational apps.

BENEFIT AND RISK

Carl Peck of the University of California, San Francisco, emphasized that in all processes, there needs to be a change in mindset or a "reset" regarding benefit-risk criteria. Benefit-risk assessments must take into account the fact that MCMs are intended for use in extreme public health emergencies (not for treatment of, for example, chronic, nonfatal conditions). Hatchett added that the development of rapid diagnostics could help facilitate the reset of benefit-risk assessments, allowing FDA to better define for whom the use of a product would outweigh potential risk (e.g., only to be used for those who test positive).

Participants discussed whether this reset of benefit-risk assessments would mean setting a level of safety that is not necessarily the same as the

TABLE 5-1 Rapid, Efficient Processes for Acquiring and Integrating Data and Predicting Favorable Benefit-Risk Ratios in Humans

Process Gaps	Regulatory Science Opportunities Cited by Panelists
Acquisition	Predictive in vitro systems (e.g., MIMIC, "liver on a chip") Bayesian study designs *In silico* systems biology and computational biology models Reset of benefit-risk criteria to be relevant to an immediate threat situation
Integration and Prediction	Bayesian, model-based integration framework Meta-modeling: integrating all data in PK/PD and/or PBPK/PD models Linking of systems biology and PBPK/PD models Mimicking adult-pediatric PK/PD dosage paradigm for animal-human prediction Reset of benefit-risk criteria to be relevant to an immediate threat situation

NOTE: PBPK/PD, physiologically based PK/PD.
SOURCE: Carl Peck. 2011. Presentation at IOM workshop; Advancing Regulatory Science for Medical Countermeasure Development.

level of safety demanded of a drug destined for the commercial market. It was noted there is a lot of precedence for benefit-risk decisions at FDA, and the agency has a strong record of making good benefit-risk decisions. The challenge for MCMs is that they are being developed for potential use in the future, for events that have not happened.

Ed Cox of CDER noted the differences between benefit-risk assessment for prophylaxis versus treatments. For a compound to be used for prophylaxis, the benefit-risk benefit calculus is complicated by the fact that while a significant number of potentially exposed people will receive the product, only a small portion may actually be at risk for the disease.

What is lacking, Peck said, is a set of processes for rapid, efficient *acquisition* and *integration* of all in vitro, animal, and human data (mechanism of action and PK/PD) that would permit *prediction* of a reasonably likely favorable clinical benefit-risk ratio in humans. Many of the cutting-edge technologies and methodologies discussed throughout the workshop could be leveraged to help close these gaps so that MCMs could be successfully approved under the Animal Rule (Table 5-1).

PLATFORMS, PROCESSES, AND TOOLS

Alan Shaw of Vaxinnate noted that FDA generally approves products or therapies on a case-by-case basis; however, throughout the workshop there was a lot of interest in platforms. FDA does not approve platforms, but tools used may be submitted for qualification.[3] The qualification process for a drug development tool is product independent. The intent is to evaluate these tools and make the data publicly available so others can use them in their development process without the need for validation every time the tool is used in association with a new product. The focus has thus far been biomarkers, patient-reported outcomes, and clinical quantitative disease progression models, but FDA is working to include animal models as well.

Biomarkers qualified by FDA for a specific context of use can then be used within this specific context use by multiple companies for multiple products. If, for example, a set of biomarkers were qualified for use in a rat model, one could now look at those in clinical studies and then, as data become available, submit another qualification package to FDA. This concept has been called "rolling qualification." A qualified tool could be considered a modular element of a platform, Rex suggested.

There was much interest in whether there could be a formalized platform qualification process. A participant explained that while the agency only approves products and not processes, a platform approach can help streamline approval of future products. For example, if there is an approved vaccine based on a vector into which appropriate genetic material was inserted, a subsequent vaccine made by inserting different genetic material into the same vector would still need to demonstrate safety and efficacy and validate manufacturing processes, but the process would presumably be faster, as much of the previous work would be relevant. The manufacturing process of the platform is, in essence, qualified (DARPA has used the term *certified*). One does not have to develop a whole new validation package. For example, the egg platform used in the manufacture of influenza vaccine is essentially a qualified manufacturing platform. As such, a new flu vaccine can be approved and manufactured within a 6-month time frame. Rex added that an adjuvant would be another example of an element or tool for which gathered data can be relevant to subsequent submissions. A participant suggested that the qualification of an animal model as being suitable to submit efficacy data could possibly be considered a platform qualification.

With regard to biomarkers, Richard Hatchett of BARDA noted that in oncology, the co-development of therapeutics and biomarkers is becoming the norm, particularly for trials of targeted therapeutics where the clinical trial population needs to be preselected based on the targeted pathway.

---

[3] See *Guidance for Industry: Qualification Process for Drug Development Tools* (Draft Guidance) http://www.fda.gov/downloads/Drugs/GuidanceComplianceRegulatoryInformation/Guidances/UCM230597.pdf (accessed June 9, 2011).

COLLABORATION, PARTNERSHIPS, AND DATA SHARING

An underlying theme in almost all regulatory science aspects of the MCM enterprise is communication (e.g., between sponsors and regulators, between funders and regulators, among stakeholders and collaborators, or from the enterprise to the public). Throughout the workshop participants were encouraged to communicate with FDA "early and often." Phyllis Arthur, Director of Healthcare Regulatory Affairs of the Biotechnology Industry Organization, emphasized the importance of communication between sponsors and FDA to establish a series of agreed-upon goals, with potentially a set of agreed-upon metrics, and accountability for all of the parties to actually achieve those goals. Ed Nuzum of NIAID said that companies should not be afraid to meet with FDA, but when they do, they need to be organized, present their data packages, and have well-formulated questions. Arthur emphasized that companies already deep in the process would benefit more from some agreed-upon accountability on both sides and more clarity and transparency as to what needs to happen along the pathway.

A challenge for small companies that are funded by BARDA is that BARDA encourages its grantees to meet with FDA to discuss moving forward with the next step, often when the company does not feel it has the data or the time necessary to prepare a meeting package. To keep to BARDA-established timelines, manufacturers are often willing to accept a certain level of risk and move forward before there is a complete dataset that could be discussed with FDA. It would be helpful, the participant said, if there could be some agreement or better clarity of the roles of the two organizations (BARDA as funder and FDA as regulator).

One approach to advancing the use of new testing methods or tools in regulatory science decision making is through precompetitive collaborative consortia involving scientists from industry, academia, and government, as well as regulators and patient representatives; a number of meeting participants expressed interest in developing such precompetitive collaborations. One example that was cited by a number of workshop participants as successful is C-Path, which, explained Marietta Anthony of C-Path, advances the development of new testing methods or tools for medical product development through establishing precompetitive collaborative consortia involving over 1,000 scientists from industry, academia, regulatory agencies, government (NIH and CDC), as well as patient representatives. Anthony explained that C-Path is a neutral, third-party entity that is able to forge partnerships and facilitate consensus development on precompetitive science between industry and regulatory authorities. Consortia activities are not product specific. Over 35 member companies in five consortia have signed a legal agreement that addresses issues of confidentiality, intellectual property, materials transfer, and antitrust. Tools/methods developed are made publicly available and are used by industry in development of medical products and by FDA in regulatory decision making. As an example, Anthony described the PSTC. The consortium includes several working groups that assess data on candidate safety biomarkers for various organs and tissues. The consortium is not focused on discovery of new biomarkers or sponsoring new research, Anthony emphasized, but on critical evaluation of existing data, conducting additional studies to fill the gaps, enhance the database, and facilitate scientific consensus. Promising biomarkers selected by the consortium are then submitted to FDA[4] and other international regulatory authorities for qualification within a specific context of use.[5]

Hatchett noted that, in addition to C-Path, the Foundation for the National Institutes of Health (FNIH) is also a successful model for collaboration between FDA, NIH, academia, and industry.

Many participants noted that data sharing across FDA centers should be increased. Nuzum suggested that government contracts and funding agreements with MCM developers should include provisions for sharing of grantee data across government agencies. The ability to share preclinical and clinical data across agencies, de-identified and pooled as appropriate, could facilitate needed meta-analyses and should be accompanied by assurances to MCM developers that their proprietary data will not be released or used to benefit a competitor. It was noted that funding initiatives from the various MCM enterprise partners (e.g., the DoD, NIH, other HHS operating divisions) may have regulatory science components, and there should be an effort to incorporate FDA input into the initial requests for applications or proposals, especially those efforts that are targeting product development.

Gail Cassell of Harvard Medical School and the Infectious Disease Research Institute in Seattle recommended that FDA forge closer ties with NIAID-funded Regional Centers of Excellence in Emerging Infections and Biodefense for access to local expertise that might be brought to bear in a public health emergency.

---

[4] See *Guidance for Industry, Qualification Process for Drug Development Tools* (Draft Guidance) http://www.fda.gov/downloads/Drugs/GuidanceComplianceRegulatoryInformation/Guidances/UCM230597.pdf (accessed June 9, 2011).

[5] Additional information is available in a special supplement to Nature Biotechnology, *The Predictive Safety Testing Consortium* 28(5):May 2010.

BOX 5-1
FDA Resource and Infrastructure Observations

Although both the National Biodefense Science Board (NBSB) and the IOM have previously issued strong recommendations for increased resources for FDA, in the current fiscal climate the significant increases that are needed are not likely to occur.

Resources are specifically needed to support the following regulatory science-related activities:

- The process of evaluating an increasing number of new technologies, and research to resolve new regulatory science challenges.
- Leadership development (scientific and professional) to strengthen science within the agency.
- Recruitment and retention of scientific talent and technical expertise.
- Establishing collaborations and partnerships.
- Academic Centers of Excellence in regulatory science to promote a better understanding of the development of MCMs and provide training.
- Information sciences.
- Biostatistics.
- Genomics.
- Synthetic biology.

SOURCE: Cassell presentation.

FDA RESOURCES AND INFRASTRUCTURE ISSUES AFFECTING REGULATORY SCIENCE

Cassell noted that many of the challenges for FDA regulatory science have roots in the agency's infrastructure issues. Cassell highlighted some of the key issues that were identified in the report *FDA Science and Mission at Risk* (FDA, 2007) as well as were mentioned throughout the workshop (Box 5-1).

Nuzum summarized the sentiment of the day, saying that FDA is doing great science and agency staff are conscientious and competent, but the tasks before them are daunting and their resources are limited and are not likely to significantly increase. As such, new paradigms are needed to find ways to work within the resources that are available.

A FRAMEWORK FOR DEFINING REGULATORY SCIENCE NEEDS

Hatchett suggested that the regulatory science needs discussed at the workshop can be sorted into tactical, operational, and strategic concerns, that is, how to deal with data in its acquisition, sharing, or management (Box 5-2).

BOX 5-2
MCM Data Issues as Tactical, Operational, and Strategic Concerns

Tactical Level—*Getting the Data*

- Developing animal models for specific applications.
- Defining appropriate models or methods to collect data (including for at-risk populations such as children, pregnant women, and others for whom absorption, distribution, metabolism, or excretion may be altered):
    - platform approaches,
    - qualification of tools (e.g., biomarkers), and
    - diagnostics.

Operational Level—*Sharing the Data*

- Cooperation, collaboration, partnerships, and sharing of data among stakeholders.
- Internal data sharing and collaboration across FDA centers.

Strategic Level—*Managing the Data*

- Maintaining competencies as new areas of science unfold (e.g., systems biology, computational biology, biostatistics).
- Benefit-risk calculus, communicating risk data.

6

Closing Remarks

Jean Hu-Primmer, acting director of FDA Medical Countermeasure Initiative Regulatory Science component, said that one of the goals of the workshop was to help identify anything else that might need to be added to the seven focus areas for Pillar 2 that FDA has defined (Box 6-1). One theme that stands out, she said, is the need for FDA to maintain the staff expertise to be able regulate new cutting-edge products. This is one of the key reasons why it is so important to maintain a robust science program inside FDA. "We strive to keep this program," she said, "despite continued limited resources, downsizing, and budget cuts." Professional development is critical, and it is important for FDA researchers to attend

BOX 6-1
Seven Regulatory Science Focus Areas of the FDA MCM Initative (Pillar 2)

1. Animal models
2. Biomarkers and clinical immunology
3. Diagnostics and devices
4. Private manufacturing and associated assay development
5. Radiation injury protection and response
6. Health and scientific computing
7. Risk communication

the same scientific seminars and symposiums as academic and industry scientists, to learn about new methods and technologies, and to connect, network, and establish collaborations. Related to this is the need to work collaboratively across government agencies and to leverage those relationships.

While Pillar 2 is just "scraping the tip of the iceberg," the MCM enterprise is moving in the right direction. The processes that have been set up for internal project proposal reviews, for example, engage all of the MCM enterprise partners. This peer-review process increases the visibility and awareness of partners, thereby creating greater potential for partnerships and collaboration with FDA and across the MCM enterprise as a whole.

The success of the FDA MCM initiative will not be dependent solely upon FDA's efforts, Hu-Primmer concluded, explaining that it depends on everyone working together, from informal collaborations to competitive grants and contracts, and all options in between.

References

ASPR (Assistant Secretary for Preparedness and Response). 2010. *The public health emergency medical countermeasures enterprise review: Transforming the enterprise to meet long-range national needs.* https://www.medicalcountermeasures.gov/documents/MCMReview Finalcover-508.pdf (accessed April 7, 2011).

Basch, E. 2010. The missing voice of patients in drug-safety reporting. *New England Journal of Medicine* 362(10):865-869.

FDA (Food and Drug Administration). 2007. *FDA science and mission at risk. Report of the Subcommittee on Science and Technology.* http://www.fda.gov/ohrms/dockets/AC/07/briefing/2007-4329b_02_01_FDA%20Report%20on%20Science%20and%20Technology.pdf (accessed April 7, 2011).

FDA. 2010. *Advancing regulatory science for public health.* http://www.fda.gov/downloads/ScienceResearch/SpecialTopics/RegulatoryScience/UCM228444.pdf (accessed April 7, 2011).

IOM (Institute of Medicine). 2010. *The Public Health Emergency Medical Countermeasures Enterprise: Innovative Strategies to Enhance Products from Discovery Through Approval: Workshop Summary.* Washington, DC: The National Academies Press.

Mandl, K. D., and I. S. Kohane. 2009. No small change for the health information economy. *New England Journal of Medicine* 360(13):1278-1281.

NBSB (National Biodefense Science Board). 2010. *Where are the countermeasures? Protecting America's health from CBRN threats.* http://www.phe.gov/Preparedness/legal/boards/nbsb/meetings/Documents/nbsb-mcmreport.pdf (accessed April 7, 2011).

NRC (National Research Council). 2010. *The prevention and treatment of missing data in clinical trials.* Washington, DC: The National Academies Press.

PCAST (The President's Council of Advisors on Science and Technology). 2010. *Report to the President on reengineering the influenza vaccine production enterprise to meet the challenges of pandemic influenza.* http://www.whitehouse.gov/sites/default/files/microsites/ostp/PCAST-Influenza-Vaccinology-Report.pdf (accessed April 18, 2011).

Poulin, P., H. M. Jones, R. D. Jones, J. W. T. Yates, C. R. Gibson, J. Y. Chien, B. J. Ring, K. K. Adkison, H. He, R. Vuppugalla, P. Marathe, V. Fischer, S. Dutta, V. K. Sinha, T. Björnsson, T. Lavé, and M. S. Ku. 2011a. PhRMA CPCDC initiative on predictive models of human pharmacokinetics, part 1: Goals, properties of the PhRMA dataset, and comparison with literature datasets. *Journal of Pharmaceutical Sciences* 100: n/a. doi: 10.1002/jps.22554

Poulin, P., R. D. O. Jones, H. M. Jones, H. M. Gibson, M. Rowland, J. Y. Chien, B. J. Ring, K. K. Adkison, M. S. Ku, H. He, R. Vuppugalla, P. Marathe, V. Fischer, S. Dutta, V. K. Sinha, T. Björnsson, T. Lavé, and J. W. T. Yates. 2011b. PHRMA CPCDC initiative on predictive models of human pharmacokinetics, part 5: Prediction of plasma concentration–time profiles in human by using the physiologically-based pharmacokinetic modeling approach. *Journal of Pharmaceutical Sciences* 100: n/a. doi: 10.1002/jps.22550.

Querec, T. D., R. S. Akondy, E. K. Lee, W. Cao, H. I. Nakaya, D. Teuwen, A. Pirani, K. Gernert, J. Deng, B. Marzolf, K. Kennedy, H. Wu, S. Bennouna, H. Oluoch, J. Miller, R. Z. Vencio, M. Mulligan, A. Aderem, R. Ahmed, and B. Pulendran. 2009. Systems biology approach predicts immunogenicity of the yellow fever vaccine in humans. *Nature Immunology* 10(1):116-125.

Ring, B. J., J. Y. Chien, K. K. Adkison, H. M. Jones, M. Rowland, R. D. Jones, J. W. T. Yates, M. S. Ku, C. R. Gibson, H. He, R. Vuppugalla, P. Marathe, V. Fischer, S. Dutta, V. K. Sinha, T. Björnsson, T. Lavé, and P. Poulin. 2011. PhRMA CPCDC initiative on predictive models of human pharmacokinetics, part 3: Comparative assessement of prediction methods of human clearance. *Journal of Pharmaceutical Sciences* 100: n/a. doi: 10.1002/jps.22552.

A

Workshop Agenda

March 29, 2011

Venable Conference Center
E11200 Capitol Room
575 7th Street, N.W.
Washington, DC 20004

Background:
During public health emergencies such as influenza pandemics or chemical, biological, radiological/nuclear (CBRN) attacks, safe and effective vaccines, drugs, diagnostics, and other medical countermeasures (MCMs) are essential to protecting national security and the well-being of the public. As highlighted in the March 2010 report of the National Biodefense Science Board, "FDA has not been able to fulfill its implicit national security mission, in large part because of a lack of resources. . . . It is imperative for America's health and progress for FDA to be provided adequate resources to bring its regulatory science into the 21st century. . . . Doing so will greatly enhance the FDA's ability to support MCM development and licensing."

In August 2010, the Department of Health and Human Services (HHS) released its *Public Health Emergency Medical Countermeasures Enterprise Review*, which made numerous recommendations to transform the public health emergency countermeasures enterprise to increase its speed, agility, capacity, and success rate, including the promotion of regulatory innovation and investment in regulatory science at FDA. To promote regulatory innovation and investment in regulatory science at FDA, FDA has established an MCM initiative. This initiative seeks to accelerate MCM development towards approval and consists of a multifaceted action plan that includes (1) enhancing the regulatory review process for the highest priority MCMs and related technologies; (2) advancing regulatory science to support MCM development and evaluation; and (3) modernizing the legal and regulatory framework to support public health preparedness and response. Regulatory science for MCM development and evaluation is essential for FDA to establish clear regulatory pathways for product approval based on the most advanced scientific foundations available and realize the promise of new technologies for flexible, rapidly scalable development and manufacturing of vaccines and other MCMs.

This workshop will (1) examine ways to advance regulatory science for MCM development and regulatory evaluation; (2) identify scientific opportunities to improve, simplify, or speed MCM development; and (3) identify tools and methods to improve the predictability and success rate of candidate MCMs.

Meeting Objectives:

- Provide a broad overview of current efforts underway at FDA and other agencies within HHS and the Department of Defense (DoD) to support the research, development, evaluation, and production of MCMs.
- Review novel scientific methodologies used by academia and industry to facilitate development of next generation vaccines, biologics, drugs, and devices.
- Identify major gaps in currently available tools to predict and evaluate product safety, efficacy, and quality.
- Identify opportunities for collaboration and coordination with FDA and among relevant federal as well as industry programs to support the MCM initiative's regulatory science program and to develop better defined pathways for product approval.
- Identify regulatory science tools and methodologies to address emerging technologies, targets, and novel products as well as innovative approaches for predicting safety and efficacy; for example, biomarkers and *in silico* modeling.

8:00 a.m. Welcome and Introductions

LESLIE BENET, *Workshop Co-Chair*
Professor, School of Pharmacy
University of California, San Francisco

JOHN REX, *Workshop Co-Chair*
Infection Clinical Vice President
AstraZeneca

8:10 a.m. Keynote Address: Importance and Promise of Regulatory Science and Charge to Workshop Participants

JESSE GOODMAN
Chief Scientist, Deputy Commissioner for Science and Public Health
Food and Drug Administration

8:25 a.m. Enterprise Activities and Needs for MCM Regulatory Science

GEORGE KORCH
Acting Principal Deputy Assistant Secretary for Preparedness and Response
Office of the Assistant Secretary for Preparedness and Response
Department of Health and Human Services

8:35 a.m. Overview of the MCM Initiative: Challenges and Opportunities

LUCIANA BORIO
Acting Director, Office of Counterterrorism & Emerging Threats
Senior Advisor for Medicine and Public Health, Office of the Chief Scientist
Office of the Commissioner, Food and Drug Administration

SESSION I: ENTERPRISE AND STAKEHOLDER PERSPECTIVES ON NEEDS TO ADVANCE MEDICAL COUNTERMEASURE REGULATORY SCIENCE

Session Objectives:

- Provide a broad overview of current efforts under way at FDA and within HHS and DoD to support the research, development, evaluation, and production of MCMs.
- Provide a broad overview of MCM development challenges and where regulatory science can advance MCM development.
- Identify and discuss the highest-priority MCM regulatory science needs to advance MCM development.

Presentation/Panel 1: FDA

8:45 a.m. Session Introduction and Panel Objectives

George Korch
Acting Principal Deputy Assistant Secretary for Preparedness and Response
Office of the Assistant Secretary for Preparedness and Response
Department of Health and Human Services

8:50 a.m. Presentations: Overview of Enterprise MCM Regulatory Science Agendas: What does the agency need?

- Identify the goals and objectives of MCM regulatory science agenda from the FDA Centers.
- Identify challenges to and information needs for MCM regulatory science programs.

Carolyn Wilson
Associate Director for Research
Center for Biologics Evaluation and Research
Food and Drug Administration

Susan McCune
Deputy Director, Office of Translational Sciences
Center for Drug Evaluation and Research
Food and Drug Administration

Murray Malin
Center for Devices and Radiological Health
Food and Drug Administration

9:50 a.m. Discussion/Question and Answer Session

10:20 a.m. BREAK

Presentation/Panel 2: Enterprise and Other Stakeholders

10:35 a.m. Panel Introduction and Objectives

Mary Pendergast
President
Pendergast Consulting

10:40 a.m. Presentations and Panel Discussion: What do the partners need?

- Identify MCM product development challenges that can be addressed with MCM regulatory science.
- Identify recurrent and overarching themes for stalled MCM product development and utilization that can be addressed through MCM regulatory science.
- Discuss highest-priority regulatory science needs to advance MCM development.

GERALD PARKER
Deputy Assistant to the Secretary of Defense
for Chemical and Biological Defense Programs
(DATSD(CBD))
Department of Defense

MICHAEL KURILLA
Director, Office of BioDefense Research Affairs
Associate Director for BioDefense Product Development
National Institute of Allergy and Infectious Disease
National Institutes of Heath

RICHARD HATCHETT
Chief Medical Officer and Deputy Director
Biomedical Advanced Research and Development Authority
Office of the Assistant Secretary for Preparedness and Response
Department of Health and Human Services

MAY CHU
Director, Laboratory Science Policy & Practice Program Office
Centers for Disease Control

RICK LYONS
Director, Infectious Disease Research Center
Chief Scientific Officer, Infectious Disease SuperCluster
Colorado State University

ERIC ROSE
CEO and Chair, Board of Directors
SIGA Technologies, Inc.

11:45 a.m. Discussion/Question and Answer Session

12:15 p.m. LUNCH

SESSION II: CUTTING-EDGE EFFORTS TO ADVANCE MEDICAL COUNTERMEASURE REGULATORY SCIENCE

Session Objectives:

- Review novel scientific methodologies used by academia and industry to discover and develop next generation vaccines, biologics, drugs, and devices.
- Identify major gaps in currently available tools to predict and evaluate product safety, efficacy, and quality.
- Identify approaches to developing regulatory science tools and methodologies to address emerging technologies, targets, and novel products as well as innovative approaches for predicting safety and efficacy.
- Discuss a path forward for the MCM regulatory science agenda.
- Discuss metrics to be used for gauging success of a scientific agenda for evaluating MCMs.
- Examine how partnerships and other collaborative approaches can facilitate the advancement and ongoing support of regulatory science for MCM development, evaluation, and utilization.

1:00 p.m. Session Introduction and Objectives

LESLIE BENET, *Workshop Co-Chair*
University of California, San Francisco

JOHN REX, *Workshop Co-Chair*
AstraZeneca

Panelists will

- Provide high-level description of MCM regulatory science tools and state of the science.
- Discuss regulatory science opportunities and challenges. What is needed to advance regulatory science to undergird regulatory decisions?
- Discuss highest-priority regulatory science needs to advance MCM development and utilization.
- Identify opportunities for partnerships and collaboration.

PANEL: EX VIVO APPROACHES TO MODELING EFFICACY

1:05 p.m. Panel Objectives and Introduction

LAUREN BLACK, *panel chair*
Senior Scientific Advisor, Navigators
Charles River Laboratories

1:10 p.m. Panel Discussion: Animal Models, *In Silico* Models, Biomarkers

***In Silico* Models**

RAMON FELCIANO
Founder & SVP of Research
Ingenuity Systems

Biomarkers

N. LEIGH ANDERSON
Founder & CEO
Plasma Proteome Institute

Animal Models

ELIZABETH LEFFEL
Director of Nonclinical Sciences
PharmAthene

1:50 p.m. Discussion with Invited Discussants and Workshop Attendees

DRUSILLA BURNS
Chief, Laboratory of Respiratory and Special Pathogens
Center for Biological Evaluation and Research

JUDY HEWITT
Chief, Biodefense Research Resources Section
Office of Biodefense Research Affairs
DMID/NIAID/NIH

VIKRAM PATEL
Office of Testing and Research
Office of Pharmaceutical Sciences, CDER

PANEL: SAFETY AND REAL-TIME MONITORING

2:25 p.m. Panel Objectives and Introduction

CARL PECK, *panel chair*
Professor, Pharmacology and Medicine
University of California, San Francisco

2:30 p.m. Panel Discussion: Post Deployment Surveillance and Side Effects; Toxicology Markers

Toxicology Markers

ROBERT HOUSE
President
DynPort Vaccine Company LLC

Post Deployment Surveillance and Side Effects

KENNETH MANDL
Faculty, Children's Hospital Informatics Program
Harvard-MIT Division of Health Sciences and Technology

2:50 p.m. Discussion with Invited Discussants and Workshop Attendees

MARIETTA ANTHONY
Predictive Safety Testing Consortium
Critical Path Institute

RICHARD FORSHEE
Associate Director for Research
Office of Biostatistics and Epidemiology
Center for Biologics Evaluation and Research

HENRY FRANCIS
Center for Drug Evaluation and Research
Food and Drug Administration

ROBERT C. NELSON
Product Safety Assurance Services, Inc.

DONNA MENDRICK
National Center for Toxicological Research
Food and Drug Administration

3:20 p.m. BREAK

PANEL: DIAGNOSTICS AND STATISTICAL TECHNIQUES

3:35 p.m. Panel Objectives and Introduction

BRUCE BURLINGTON, *panel chair*
Independent Consultant

3:40 p.m. Panel Discussion: Diagnostics

Diagnostics

LT. COL. DANIEL WATTENDORF
U.S. Air Force
Program Manager, Defense Sciences Office
DARPA

3:50 p.m. Discussion with Invited Discussants and Workshop Attendees

CHARLES DAITCH
Chief Executive Officer
Akonni Biosystems

DAVID ECKER
Founder, DVP and Carlsbad General Manager
Ibis Biosciences

SALLY HOJVAT
Center for Diagnostics and Radiological Health, FDA

4:25 p.m. Panel Discussion: Statistical Techniques

Statistical Techniques

STEPHEN RUBERG
Distinguished Research Fellow & Scientific Leader, Advanced Analytics
Eli Lilly & Co.

4:35 p.m. Discussion with Invited Discussants and Workshop Attendees

ESTELLE RUSSEK-COHEN
Acting Division Director, Division of Biostatistics
Office of Biostatistics and Epidemiology, CBER/FDA

JEFFREY WETHERINGTON
Research Director, Statistics
GlaxoSmithKline

5:00 p.m. ADJOURN

March 30, 2011

Venable Conference Center
E11200 Capitol Room
575 7th Street, N.W.
Washington, DC 20004

SESSION II, CONT'D: CUTTING-EDGE EFFORTS TO ADVANCE MEDICAL COUNTERMEASURE REGULATORY SCIENCE

Session Objectives:

- Review novel scientific methodologies used by academia and industry to discover and develop next generation vaccines, biologics, drugs, and devices.
- Identify major gaps in currently available tools to predict and evaluate product safety, efficacy, and quality.
- Identify approaches to developing regulatory science tools and methodologies to address emerging technologies, targets, and novel products as well as innovative approaches for predicting safety and efficacy.
- Discuss a path forward for the MCM regulatory science agenda.
- Discuss metrics to be used for gauging success of a scientific agenda for evaluating MCMs.
- Examine how partnerships and other collaborative approaches can facilitate the advancement and ongoing support of regulatory science for MCM development, evaluation, and utilization.

8:00 a.m. Welcome and Day 2 Overview

Leslie Benet, *Workshop Co-Chair*
University of California, San Francisco

John Rex, *Workshop Co-Chair*
AstraZeneca

Panel: Antimicrobials, Vaccines, and Vaccine Adjuvants

8:15 a.m. Panel Objectives and Introduction

Linda A. Miller, *panel chair*
Director, Clinical Microbiology
GlaxoSmithKline

8:20 a.m. Panel Discussion: Antimicrobials, Vaccines, and Vaccine Adjuvants

Antimicrobials

Kevin Judice
CEO & Chief Scientific Officer
Achaogen

Vaccines

Alan Magill
Director, Division of Experimental Therapeutics
Walter Reed Army Institute of Research

Adjuvants

Debbie Drane
Senior VP R&D
CSL Biotherapies

9:00 a.m. Discussion with Invited Discussants and Workshop Attendees

Ed Cox
Center for Drug Evaluation and Research
Food and Drug Administration

DAVID FRUCHT
Capt., U.S. Public Health Service
Chief, Laboratory of Cell Biology
Division of Monoclonal Antibodies
CDER/FDA

BASIL GOLDING
Center for Biologics Evaluation and Research

HANA GOLDING
Chief, Laboratory of Retrovirus Research, Division of Viral Products
Office of Vaccines Research and Review, CBER

ED NUZUM
Chief, Biodefense Vaccines & Other Biological Products Development Section
Office of Biodefense Research Affairs
DMID/NIAID/NIH

ALAN SHAW
Chief Scientific Officer
Vaxinnate

PANEL: THE FUTURE

9:45 a.m. Panel Objectives and Introduction

GIGI KWIK GRONVALL, *panel chair*
Senior Associate, Center for Biosecurity
UPMC

9:50 a.m. Panel Discussion: Synthetic Biology, Computational Biology, and Platform Technologies

Synthetic Biology

JOHN GLASS
Senior Scientist
J. Craig Venter Institute

Computational Biology

MICHAEL KATZE
Professor of Microbiology
University of Washington

Platform Technology

PAT IVERSEN
Senior Vice President of Research & Innovation
AVI BioPharma

10:30 a.m. Discussion with Invited Discussants and Workshop Attendees

WILLIAM FOGLER
Senior Director, Portfolio Analysis and Planning
Intrexon Corporation

DONNA MENDRICK
Director, Division of Systems Biology
National Center for Toxicological Research
Food and Drug Administration

HARVEY RUBIN
Executive Director, Institute for Strategic Threat Analysis and Response
University of Pennsylvania

11:00 a.m. BREAK

SESSION III: MEDICAL COUNTERMEASURE REGULATORY SCIENCE NEEDS FOR AT-RISK POPULATIONS

Session Objectives:

- Identify needs that are specific or unique to at-risk populations (e.g., pediatric populations) that should be considered in developing a regulatory science agenda for MCM development.
- Provide an overview of where regulatory science can advance MCM development for these populations.
- Discuss a MCM regulatory science agenda for at-risk populations.

Panelists will

- Identify recurrent and overarching challenges for MCM product development, implementation, and use specific to at-risk populations that can be addressed with MCM regulatory science.
- Discuss highest-priority regulatory science needs to advance MCM development and utilization for at-risk populations. What tools could be used to support regulatory review and determinations of use in these populations in an emergency?
  - Discuss issues for at-risk populations such as: lack of dosing information, and needs for safety and efficacy models.
  - Specific scientific gaps in treating or vaccinating pregnant women.

11:15 a.m. Introduction and Session Objectives

JOHN BRADLEY, *panel chair*
Director, Infectious Diseases
Rady Children's Hospital San Diego

11:20 a.m. Panel Discussion

Pediatric

DAVID SIEGEL
NICHD

Pregnancy/OB-Gyn

CHRISTIAN R. MACEDONIA
Joint Chiefs of Staff, the Pentagon

11:45 a.m. Discussion with Attendees

PRAVIN JADHAV
Center for Drug Evaluation and Research
Food and Drug Administration

NANCY MESSONNIER
Centers for Disease Control

LISA MATHIS
Center for Drug Evaluation and Research
Food and Drug Administration

ROBERT NELSON
Senior Pediatric Ethicist/Lead Medical Officer
Office of Pediatric Therapeutics
Office of the Commissioner, Food and Drug Administration

12:30 p.m. LUNCH

SESSION IV: FUTURE DIRECTIONS—DISCUSSION WITH WORKSHOP PARTICIPANTS AND ATTENDEES

Session Objectives: Discuss what opportunities and challenges exist to implementing the models discussed at the workshop for advancing regulatory science. What should be on the agenda for regulatory science for development and evaluation of MCMs? Discuss strategies and needs to implement the MCM regulatory science agenda.

LESLIE BENET, *Workshop Co-Chair*
University of California, San Francisco

JOHN REX, *Workshop Co-Chair*
AstraZeneca

1:30 p.m. Discussion with Panelists and Workshop Attendees led by workshop co-chairs

- Synthesize workshop discussions.
- Propose key opportunities to develop a research agenda and roadmap for the MCM regulatory science initiative.

PHYLLIS ARTHUR
Director, Health & Regulatory Affairs
Biotechnology Industry Organization

GAIL CASSELL, *Drug Forum Co-Chair*
Visiting Professor, Department of Social Medicine, Harvard Medical School
Vice President, TB Drug Discovery, Infectious Disease Research Institute, Seattle

RICHARD HATCHETT
Chief Medical Officer and Deputy Director
Biomedical Advanced Research and Development Authority
Office of the Assistant Secretary for Preparedness and Response
Department of Health and Human Services

CARL PECK
Professor, Pharmacology and Medicine
University of California, San Francisco

ED NUZUM
Chief, Biodefense Vaccines & Other Biological Products Development Section
Office of Biodefense Research Affairs
DMID/NIAID/NIH

ERIC ROSE
CEO and Chair, Board of Directors
SIGA Technologies, Inc.

3:00 p.m. Discussion with Workshop Attendees led by workshop co-chairs

LESLIE BENET, *Workshop Co-Chair*
University of California, San Francisco

JOHN REX, *Workshop Co-Chair*
AstraZeneca

4:30 p.m. Closing Remarks

JEAN HU-PRIMMER
Senior Advisor for Regulatory Policy
Acting Director, MCMi Regulatory Science
Office of the Chief Scientist
Office of the Commissioner, Food and Drug Administration

B

Participant Biographies

N. Leigh Anderson, Ph.D., is Founder and CEO of the Plasma Proteome Institute, Washington D.C. (www.plasmaproteome.org). The institute aims to foster a comprehensive exploration of the proteins of human blood plasma (the plasma proteome), improved quantitation of potential disease markers, and the rapid application of novel protein measurements in clinical diagnostics. Dr. Anderson obtained his B.A. in Physics with honors from Yale and a Ph.D. in Molecular Biology from Cambridge University (England) where he worked with M. F. Perutz as a Churchill Fellow at the MRC Laboratory of Molecular Biology. Subsequently he founded (with Dr. Norman Anderson) the Molecular Anatomy Program at the Argonne National Laboratory (Chicago) where his work in the development of 2-D electrophoresis and molecular database technology earned him, among other distinctions, the 1983 Pittsburgh Analytical Chemistry Award. Prior to founding PPI, Dr. Anderson was Chief Scientific Officer at Large Scale Biology Corporation, whose proteomics division he founded in 1985, and co-led a successful Nasdaq IPO based largely on the proteomics technology platform. More recently Dr. Anderson has developed novel technologies for quantitation of protein biomarkers using mass spectrometry, receiving the 2009 HUPO Distinguished Achievement Award in Proteomic Science. Dr. Anderson currently serves as a Principal of Anderson Forschung Group LLC, a member of the Board of Directors of Luna Innovations (a developer of novel sensors and materials), associate editor of the journal *Clinical Chemistry*, and sits on numerous scientific advisory boards. Dr. Anderson has published more than 150 scientific papers, one book, and 32 patents.

Marietta Anthony, Ph.D., is at the Critical Path Institute (C-Path), which builds collaborative partnerships in a new model of drug development and in support of FDA's regulatory mission. She is the associate director of the Arizona Center for Education & Research on Therapeutics (CERT), which focuses on drug safety. Dr. Anthony was a senior health policy analyst at three federal agencies—the Agency for Healthcare Policy and Research, the Food and Drug Administration, and the National Institutes of Health. Additionally, Dr. Anthony was in the Department of Pharmacology at Georgetown University and later, Vice President of Health Sciences in Women's Health the University of Arizona. Dr. Anthony served on an IOM panel on women's health research.

Phyllis Arthur, M.B.A., joined the Biotechnology Industry Organization (BIO) in July 2009 as the Director of Healthcare Regulatory Affairs. In this role Ms. Arthur is responsible for working with member companies in vaccines and biodefense on policy, legislative, and regulatory issues. Prior to joining BIO, Ms. Arthur worked in numerous marketing and sales positions for Merck & Co Inc. in their Vaccine Division. Over her 16-year career in vaccines, Ms. Arthur launched several exciting new vaccines in the United States and internationally, worked closely with clinical and academic thought leaders in infectious diseases and oncology, and ran a large sales organization of over 75 representatives and managers. Before graduate school, Ms. Arthur worked as a research assistant for two economists at the Brookings Institution in Washington, D.C. There she conducted economic analyses related to savings and investment policies for the OECD countries. Ms. Arthur received her B.A. in 1987 in Economics and International Politics from Goucher College and her M.B.A. in 1991 from the Wharton School of Business at the University of Pennsylvania.

Leslie Z. Benet, Ph.D., Professor, Department of Biopharmaceutical Sciences, UCSF, has received honorary doctorates from six universities: Uppsala (1987), Leiden (1995), Illinois at Chicago and PCP&S (1997), Long Island (1999), and Athens (2005). During 1986, Dr. Benet founded and became the first president of the American Association of Pharmaceutical Scientists (AAPS). Elected to IOM membership in 1987, he has chaired the following committees: Clinical Applications of Mifepristone RU486 and Other Antiprogestins; Pharmacokinetics and Drug Interactions in the Elderly and Special Issues in Elderly African-American Populations; and Accelerating the Research, Development, and Acquisition of Medical Countermeasures Against Biological Warfare Agents. His many awards include AAPS Distinguished Pharmaceutical Scientist (1989); American Association of Colleges of Pharmacy Volwiler Research Achievement (1991); American Society for Clinical Pharmacology and Therapeutics Rawls-Palmer Progress in Medicine Award (1995); American Pharmaceutical Association Higuchi Research Prize (2000); AAPS Wurster Award in Pharmaceutics (2000); International Pharmaceutical Federation Høst-Madsen Medal (2001); Pharmaceutical Sciences World Congress Research Achievement award; and Controlled Release Society Career Achievement in Oral Drug Delivery (2004). His research interests, more than 470 publications, and 11 patents are in the areas of pharmacokinetics, biopharmaceutics, drug delivery, and pharmacodynamics. He is listed among the 250 most cited pharmacologists worldwide and is in the top 5 percent in NIH research funding over the past 25 years.

Lauren E. Black, Ph.D., is employed by Charles River Laboratories (CR) as a Senior Scientific Advisor in CR's Navigator Services. She left FDA and now consults internationally on drug program strategy, biologics, translational research, and safety programs. She has over 20 years' experience in drug development, clinical risk mitigation, regulatory negotiations, and strategic planning. Dr. Black graduated from Carnegie Mellon University and studied physiologic responses to opiates/shock. With a doctorate in Pharmacology and Toxicology from the VCU School of Medicine, she did her postdoctoral work at NIH/NINDS, publishing in the area of dopamine receptor regulation. In 1991, Dr. Black went to FDA/CDER (Drugs) as a Reviewing Pharmacologist and reviewed over 40 INDs. Her reviews supported NDA approvals for Aldara, Abreva, Flumadine, Prograf, CellCept, Neoral, and Rapamune. These latter drugs include the primary immunosuppressants still used in transplantation. She worked on committees producing FDA guidance on immunotoxicology, rheumatoid arthritis, xenotransplantation, biologics, and oligonucleotides. She represented FDA views on risk assessment and safety programs at scientific conferences. In 1995, Lauren transferred to FDA/CBER (Biologics) where she reviewed protein drugs for chronic diseases such as Crohn's, RA (Humira), MS (Tysabri), and psoriasis (Amevive). She promoted the use of homolog monoclonals for nonclinical safety evaluation. This novel safety program approach supported the marketing approvals for Remicade, Raptiva, and Cimzia. She also reviewed several monoclonals that caused cytokine release in patients. Dr. Black was invited to brief the Dermatologic Advisory Committee on risk mitigation employing PK/PD-based dose regimens. She co-led the committee that produced the FDA guidance on first-in-human starting dose. These efforts presaged the later CHMP guidance on high-risk therapeutics. In CBER, Dr. Black reviewed over 400 INDs and eight marketing applications, including Carticel, the first FDA-approved cell/regenerative medicine product. She also reviewed bone marrow, human, and xenogeneic cell-based products for severe and life-threatening diseases of both adults and children. She addressed the Advisory Panel on xenotransplantation on preclinical support for clinical trials. Dr. Black participated in working groups on Tissue Engineering and co-authored the nonclinical section of the CBER guidance on xenotransplantation products. This document remains the only FDA guidance pertaining to nonclinical evaluation of cellular products. Dr. Black has subsequently published and taught in the areas of risk assessment, protein and cell therapy, translational medicine, and immunotoxicology. She is a full member of the Society of Toxicology. Additionally, she serves on advisory boards for pharma and NIH. Dr. Black is Chair of the Special Biologics-Expert Working Group, a BIO/BioSAFE committee. She is a CR delegate to the International Life Sciences Institute/HESI and a member of the NC3Rs committee on nonhuman primate use in biologics development. Dr. Black owns no stock in any pharmaceutical development firms.

Luciana Borio, M.D., is a Senior Advisor for Medicine and Public Health in the Office of the Chief Scientist, U.S. Food and Drug Administration (FDA) and Acting Director of the Office of Counterterrorism and Emerging Threats, where she is leading the implementation of FDA's Medical Countermeasures Initiative. Dr. Borio joined the FDA in 2008 as a medical reviewer in the Office of Vaccine Research and Review, Center for Biologics Evaluation and Research. Prior to joining the FDA, Dr. Borio was a Senior Associate at the Center for Biosecurity of UPMC and Assistant Professor of Medicine at the University of Pittsburgh from 2003–2008, where she worked to develop policies to improve the nation's preparedness for bioterrorism and to support threat assessments, medical countermeasure development, and medical response plans. Prior to that, she was a Senior Fellow at the Johns Hopkins University Center for Civilian Biodefense Strategies and Assistant Professor of Medicine in the Division of Infectious Diseases at Johns Hopkins University. Dr. Borio served at the U.S. Department of Health and Human Services (HHS) as an Advisor on Biodefense Programs from 2001–2008. At HHS, she implemented and managed mathematical modeling projects to assess the health effects of bioterrorism on civilians and to inform medical countermeasure procurement activities for the Office of Preparedness and Response. Dr. Borio is an infectious disease physician and continues to practice medicine at Johns Hopkins Hospital. She serves on the Pandemic Influenza Task Force of the Infectious Diseases Society of America. She has previously served on their Global and Public Health Committee and on the National Research Council's committee on Methodological Improvements to the

Department of Homeland Security's Biological Agent Risk Analysis. She has lectured and published extensively on infectious diseases and biodefense. Dr. Borio is a member of the Infectious Diseases Society of America. Dr. Borio received a B.S. in 1992 and an M.D. in 1996 from the George Washington University. She completed residency in 1999 in Internal Medicine at the New York Presbyterian Hospital-Cornell Medical Center, and subsequently completed a combined fellowship in Infectious Diseases (at Johns Hopkins University) and Critical Care Medicine (at the National Institutes of Health).

Dr. John Bradley, M.D., received his pediatric infectious diseases training at Stanford University. He has focused his research on the clinical aspects of infectious diseases with antibacterial, antiviral, and antifungal agents. He served on the American Academy of Pediatrics Committee on Infectious Diseases from 2004–2010, helping to create national policies for infectious diseases in pediatrics. He is currently on the Pediatric Infectious Diseases Society Council. Dr. Bradley is also currently the Chair of the AAP/IDSA/PIDS/ATS Writing Group for Guidelines for Pediatric Community-Acquired Pneumonia. He is on the FDA's Advisors and Consultants Staff, having served for six years on the AntiInfective Drug Advisory Committee. He sits on the Infectious Diseases Society's Task Force on Antimicrobial Drug Availability, whose charge is to bring together members of Congress, FDA, and the pharmaceutical industry to facilitate antibiotic discovery and approval for antibiotic-resistant bacterial pathogens. Representing IDSA, he is a member of the FNIH Biomarkers Consortium, charged with assisting develop current clinical trial designs for ABSSSI, CABP that are acceptable to FDA, academia, and the pharmaceutical industry. He has testified on behalf of the American Academy of Pediatrics before the House Subcommittee on Health (Waxman) on the need for new antibacterial agents for children. He has been the Director of the Division of Infectious Diseases at Rady Children's Hospital–San Diego for the past 22 years, and is Chief, Division of Infectious Diseases, Department of Pediatrics, at the University of California, San Diego School of Medicine.

D. Bruce Burlington, M.D., an infectious disease internist, is a well-known independent consultant on pharmaceutical product development and regulatory affairs. He has special interests in helping companies plan development of their drugs based on FDA and European Union requirements; prepare for meetings with FDA or its advisory committees; develop risk management plans; conduct product due diligence evaluations; and set up process, organizations, and staffing plans to achieve their regulatory obligations. He has been a senior executive in both FDA and the pharmaceutical industry. He blends long experience in development strategy with insightful analysis of the underlying medical problems, patient needs, how the results will be viewed by FDA and EMEA, and what outcomes will result in commercial success. Dr. Burlington was Executive Vice President and worldwide head of Regulatory Affairs, Human Safety, and Quality at Wyeth. He led the company in the development and U.S. and global registration of many products as well as improving Wyeth's compliance posture. He also successfully navigated the company through an FDA consent degree. During these 8 years, as a member of many Wyeth governance councils and committees, including the executive licensing, capitol expenditure, and commercial councils, he participated broadly and in depth analyzing the complex business forces driving industry. Before joining Wyeth, Dr. Burlington served at FDA for 17 years. He was the first physician named as director at of the Center for Devices and Radiological Health (CDRH) where he led major changes, increased the rigor of clinical investigation for medical devices, and championed innovations in how the center could work more productively with industry. Before that he was a research immunologist and then a manager in both the Center for Biologics Evaluation and Research (CBER) and the Center for Drug Evaluation and Research (CDER). In these centers he had responsibility for viral vaccines, investigational biologics, BLA review, NDA approvals, and generic drugs. As medical Deputy Director in CDER (acting), he also oversaw policy and compliance decisions for pharmaceuticals. Dr. Burlington is a frequent public speaker on drug development, risk management, and how to work successfully with regulators. He has organized and chaired numerous symposia and courses in the field including the American Course in Drug Development and Regulatory Science, and has been a member of several trade association and public committees as well as four boards of directors (currently AstraZeneca and Cangene) and three pharmaceutical company scientific advisory boards.

Drusilla Burns, Ph.D., graduated from Tulane University with a major in chemistry. She received her Ph.D. in biochemistry from the University of California, Berkeley, after which she completed a postdoctoral fellowship at the National Institutes of Health. She then joined the Center for Biologics Evaluation and Research, FDA, where she is currently Chief of the Laboratory of Respiratory and Special Pathogens. Her research focuses on microbial pathogenesis, host response, and vaccines against bacterial diseases. She has served on the editorial boards of *Infection and Immunity* and the *Journal of Biological Chemistry*, and has served as an Editor of *Infection and Immunity*. At FDA, she is involved in the regulation of anthrax, pertussis, and other bacterial vaccines.

Gail H. Cassell, Ph.D., is a Visiting Professor in the Department of Social Medicine, Harvard Medical School, and Vice President of TB Drug Discovery of the not-for-profit Infectious Disease Research Institute in Seattle. Dr. Cassell has recently retired as Vice President, Scientific Affairs and Distinguished Lilly Research Scholar for Infectious Diseases, Eli Lilly and Company in Indianapolis, Indiana. She is former Charles H. McCauley Professor and Chair of the Department of Microbiology, University of Alabama Schools of Medicine and Dentistry at Birmingham, a department that ranked first in research funding from the National Institutes of Health (NIH) during the decade of her leadership. She obtained her B.S. from the University of Alabama in Tuscaloosa and in 1993 was selected as one of the top 31 female graduates of the twentieth century. She obtained her Ph.D. in microbiology from the University of Alabama at Birmingham and was selected as its 2003 Distinguished Alumnus. She is past President of the American Society for Microbiology (the oldest and single largest life sciences organization, with a membership of more than 42,000). She was a member of the NIH Director's Advisory Committee and of the Advisory Council of the National Institute of Allergy and Infectious Diseases. She was named to the original Board of Scientific Councilors of the Center for Infectious Diseases, Centers for Disease Control and Prevention (CDC), and served as chair of the board. She recently served a 3-year term on the advisory board of the Director of CDC and as a member of the Secretary of Health and Human Services' Advisory Council of Public Health Preparedness. Currently she is a member of the FDA Science Board. Since 1996 she has been a member of the U.S.–Japan Cooperative Medical Science Program, responsible for advising the respective governments (U.S. State Department/Japanese Ministry of Foreign Affairs) on joint research agendas. She has served on several editorial boards of scientific journals and has authored more than 250 articles and book chapters. Dr. Cassell has received national and international awards and an honorary degree for her research in infectious diseases. She is a member of the IOM. Dr. Cassell has been intimately involved in the formulation of science policy and legislation related to biomedical research and public health. For 9 years she was chair of the Public and Scientific Affairs Board of the American Society for Microbiology; she has served as an advisor on infectious diseases and indirect costs of research to the White House Office of Science and Technology Policy, and has been an invited participant in numerous congressional hearings and briefings related to infectious diseases, antimicrobial resistance, and biomedical research. She has served two terms on the Liaison Committee on Medical Education (LCME), the accrediting body for U.S. medical schools, as well as other national committees involved in establishing policies on training in the biomedical sciences.

She recently completed a term on the Leadership Council of the School of Public Health of Harvard University. Currently she is a member of the Executive Committee of the Board of Visitors of Columbia University School of Medicine, the Executive Committee of the Board of Directors of the Burroughs Wellcome Fund, Research!America, and the Advisory Council of the Johns Hopkins School of Nursing.

May C. Chu, Ph.D., is the Director of the Laboratory Science Policy and Practice Program Office (LSPPPO). Dr. Chu draws upon her public health laboratory, applied research experiences as well as her laboratory program policy and practice experience to lead the LSPPPO. Dr. Chu began her career with CDC in the Division of Vector-Borne Infectious Diseases in Ft. Collins, Colorado as a Research Microbiologist analyzing the molecular epidemiology of dengue viruses and continued there as the chief of the Bacterial Zoonotic Diseases Diagnostic and Reference Section focusing on diagnostic/applied research on plague and tularemia and supporting diagnostic research work on borreliosis. Just before joining LSPPPO in November 2010, Dr. Chu completed 6.5 years service at the World Health Organization (WHO) where she served as team leader for Laboratory Alliances and Biosafety in the International Health Regulations Coordination Department and briefly served as the Associate Director for Laboratory Science in the Division of Preparedness and Emerging Infections, National Center for Emerging and Zoonotic Infectious Diseases. LSPPPO is positioned to champion for quality laboratory services serving the public health agenda through facilitating, coordination and supporting cross-cutting laboratory functions that impact policy and practice. A facilitated global discussion on the direction of how laboratory services should look like by "2020 and beyond" is critical in austere times against a background of rapidly evolving technological advances. The resources within LSPPPO and the strong partnerships with CDC colleagues and external stakeholders are critical to this effort. Dr. Chu serves as the Chair of the International Board of the American Society for Microbiology that focuses on promoting microbiology and to connect the microbiologists. Dr. Chu received her B.S. from Michigan State University majoring in Microbiology and Public Health. She received her doctorate degree in Tropical Medicine and Medical Microbiology from the John A. Burns School of Medicine, University of Hawaii at Manoa.

Ed Cox, M.D., M.P.H., is currently the Director of the Office of Antimicrobial Products within the Office of New Drugs, Center for Drug Evaluation and Research, U.S. Food and Drug Administration. Dr. Cox received his undergraduate degree in chemistry from the University of North Carolina at Chapel Hill and his medical degree from the University of North

Carolina School of Medicine. He completed an internship and residency in Internal Medicine at the Hospital of the University of Pennsylvania in Philadelphia. He went on to complete a fellowship in Infectious Diseases at the National Institute of Allergy and Infectious Diseases at the NIH in Bethesda, Maryland. Dr. Cox is board certified in Internal Medicine and Infectious Diseases.

Charles Daitch, Ph.D., is Founder and President, Akonni Biosystems. Dr. Daitch has 20 years experience encompassing a broad range of disciplines including chemistry, biology, biosensors and chemical/biological defense. He directs multidisciplinary teams of scientists focused on the development of infectious disease diagnostic tools that implement novel sample preparation, molecular recognition, and optical-based detection technologies. Dr. Daitch has significant experience in sensor system product development employing (1) automated miniaturized biomolecule purification and signal amplification in complex samples, and (2) integration of such microfabricated components as microarrays, thermal electric control, microfluidic fluid flow, and electro-optical detection. He brings strong product development and R&D experience from the NIH/FDA, the USDA, and Sandia National Labs. Dr. Daitch was recruited by HandyLab Inc., a microfluidics diagnostics company, to serve on the executive management team as Vice President of R&D. In this role, he assisted in strategic planning, business alliances and agreements, and R&D collaborations. Prior, Dr. Daitch launched a biodefense R&D business unit for Veridian Corporation, a premier defense contractor, and led the growth of the operation to 24 employees and $5M in annual revenue. He has four U.S. patents, 17 peer-reviewed publications, and has served as principle investigator on over $12 million in government grants and contracts.

Debbie Drane, B.Sc., has been with CSL in the R&D Division for over 25 years. During this time she has had a number of roles and has substantial experience in most areas of vaccine development. In particular she has worked on CSL's proprietary ISCOMATRIX adjuvant for many years in both technical and management positions, playing a key role in the successful development of the technology. As the SVP R&D reporting to the CSO, Ms. Drane is responsible for all activities related to the ISCOMATRIX adjuvant technology including research, manufacturing and commerialization as well as maintaining successful relationships with major partners of the technology such as Merck and Pfizer.

David J. Ecker, Ph.D., is a Divisional Vice President and a General Manager at Abbott. Dr. Ecker was the founder of Ibis Biosciences, now a subsidiary of Abbott Molecular, Inc., and was a co-founder of Isis Pharmaceuticals. Along with his core team, he was a primary inventor of the Ibis technology and Ibis T5000 (formerly TIGER) technology, now commercially marketed as the Abbott PLEX-ID. The technology was developed for infection control, infectious disease diagnostics, and biological weapons defense and human and microbial forensics, sponsored by DARPA, CDC, NIAID, FBI, DHS, and other U.S. government agencies. He is currently responsible for Abbott's Ibis site in Carlsbad, California, which has approximately 80 employees. He is responsible for the science, patents, business development, strategic direction, and the management of corporate and government partnerships. Dr. Ecker has over 28 years of experience in the pharmaceutical and biotechnology industry in drug discovery and diagnostic platform technology development. Dr. Ecker received a B.A. in Biology and Chemistry from the College of New Jersey, NJ, and his Ph.D. in Biochemistry from Utah State University, Utah.

Ramon Felciano, Ph.D., is the Founder and SVP of Research, Ingenuity Systems. Dr. Ramon Felciano was born in San Francisco, California. He holds a Ph.D. and M.S. in Biomedical Informatics, a B.S. in Computer Science, and a B.A. in English and French Literature from Stanford University. While at Stanford, Dr. Felciano performed research on semi-automated methods for designing intelligent user interfaces; scientific information visualization; and distributed, knowledge-based biomedical information systems. His doctoral research (with Dr. Russ Altman) focused on automatic generation of biomedical graphics and their use as the bases for biomedical user interfaces. Dr. Felciano is a founding member of the RiboWeb, a seminal project to build a World Wide Web-based knowledge base to support collaborative molecular biology over the Internet. Dr. Felciano's other research efforts include a patented user-tracking technology for the World Wide Web; and a formal study of Human Error in Medicine and its impact on the design of biomedical information systems. Dr. Felciano co-founded Ingenuity in 1998 to improve human health by increasing research productivity in the scientific enterprise. Dr. Felciano leads the company's strategic R&D and collaborations in countermeasures research; systems biology; and predictive analytics for drug and biomarker discovery, large-scale scientific data integration, scientific drug discovery services, and research informatics for distributed drug discovery and development. The resulting scientific tools deliver systems biology expertise to biologists, chemists, clinical researchers, and informatics specialists in global pharmaceutical R&D organizations as well as government, academic, and not-for-profit research institutions. Ingenuity's technology and discovery approach has been validated by tens of thousands of global researchers that have successfully applied the

Ingenuity platform and discovery approach across all major therapeutic research areas to improve discovery insights and speed time to market for drugs and diagnostics. Prior to founding Ingenuity, Dr. Felciano co-founded SUMMIT, the Stanford University Medical Media and Information Technologies lab, where he held the position of Associate Director for 4 years, and Digital Alchemy, a strategic research and design consultancy based in San Francisco, California.

William E. Fogler, Ph.D., has extensive experience in translational research with a proven track record of designing and managing multiple therapeutic projects from discovery through early clinical evaluation. His insightful understanding of applied therapeutics was honed during a decade-long tenure as an innovative investigator at the National Institutes of Health during which he held a dual appointment at the National Cancer Institute's Cancer Therapy Evaluation Program and Laboratory of Experimental Immunology. He joined Intrexon from EntreMed, Inc., of Rockville, Maryland, where he served as Senior Director of Translational Research. Dr. Fogler received his Ph.D. in Pathology from the University of Maryland at Baltimore, School of Medicine; his M.S. in Biomedical Science from Hood College in Frederick, Maryland; and his B.S. in Biology from the University of Maryland at College Park. He is an accomplished scientist-inventor with 68 peer-reviewed publications, 16 book chapters, and 15 patents to date.

Richard Forshee, Ph.D., is the Associate Director for Research for the Office of Biostatistics and Epidemiology at the Center for Biologics Evaluation and Research in the U.S. Food and Drug Administration. He provides leadership and support across a range of research projects on genomics, bioinformatics, clinical trial research, and other areas. Previously, Dr. Forshee developed quantitative risk assessment models to improve the understanding of the likely public health impact of risk management options for biologics products, such as blood products, vaccines, and human cell and tissue products. He has recently given presentations to the Blood Products Advisory Committee on risk assessments of selective testing strategies for *Trypanosoma cruzi* and on the public health impact of hypothetical home-use HIV tests with different performance characteristics. Dr. Forshee has published numerous scientific articles on public health issues. Before joining FDA, he was a Research Associate Professor and the Director of the Center for Food, Nutrition, and Agriculture Policy at the University of Maryland, College Park.

Henry L. Francis, M.D., is Deputy Office Director, Office of Surveillance and Epidemiology, at FDA's Center for Drug Evaluation and Research

Food (CDER). In October 2007, Dr. Francis joined the Office of Surveillance and Epidemiology (OSE) in CDER. Dr. Francis works with the OSE director to lead five divisions of pharmacy and clinical scientists in the detection and study of adverse medical events occurring after the release of new drugs into the American health market, also called the postmarket period. Dr. Francis's specific interest is in the development of data-mining techniques to enhance pharmacovigilance capabilities in national medication use and health care databases. Prior to working in FDA, Dr. Francis was a basic and clinical researcher in the National Institute of Allergy and Infectious Diseases. As an AIDS clinical investigator, he worked in several clinical and epidemiologic research projects conducting AIDS and tropical research projects in the Democratic Republic of the Congo (DRC, formerly known as Zaire) and other projects in the Caribbean and the South Pacific. In the DRC, Dr. Francis was the Director of the U.S. Public Health Service & Belgian Project SIDA (AIDS research) Research Laboratories in Kinshasa. Dr. Francis served as the first Director of the National Institute on Drug Abuse's (NIDA) Center on AIDS and Other Medical Consequences of Drug Abuse (CAMCODA). CAMCODA's mission was to establish sustainable AIDS-specific research projects in coordination with the other NIDA projects investigating drug abuse prevention and treatment. As a clinician, Dr. Francis was an assistant professor of medicine at the Johns Hopkins University School of Medicine's Division of Infectious Diseases where he served as the Principal Medical Officer of the Broadway Women's Drug Use Center and as the Ryan White Title III investigator and Medical Director of the Baltimore City Department of Health's Sexually Transmitted Diseases clinics. Dr. Francis was born in Waterbury, Connecticut, is a graduate of Amherst College, and received his M.D. from the Howard University College of Medicine in Washington, D.C. He completed his Internal Medicine residency training at the Long Beach Veterans Administration Hospital in California and his infectious diseases specialty training at the Johns Hopkins Hospital Division of Infectious Diseases, Baltimore, Maryland.

David M. Frucht, M.D., was born in Maryland. He attended the University of Virginia, receiving an interdisciplinary B.A. degree in the Echols Scholars Program in 1986. He attended Duke University for medical school and internal medicine residency training, which he completed in 1993. He then undertook infectious diseases training at NIAID/NIH (1993–1997), while conducting research on patients with immunodeficiencies in the laboratory of Dr. Steven Holland. From 1997 to 2001, he pursued a second research fellowship program studying mechanisms of signal transduction in the laboratory of Dr. John O'Shea. Subsequently, he joined the

Division of Monoclonal Antibodies as a principal investigator where he has studied the mechanisms of action of anthrax lethal toxin. Dr. Frucht's laboratory has published numerous original research papers and review articles related to anthrax toxin. Over the past year he was awarded two U.S. patents for the development of new anthrax toxin bioassays. Since 2008, Dr. Frucht has served as the Chief of the Laboratory of Cell Biology. In this capacity, he has either primary or supervisory oversight for nearly all monoclonal antibodies used as MCMs in the United States. In addition to his routine duties, Dr. Frucht is a commissioned officer in the U.S. Public Health Service, having deployed on several disaster missions including Hurricane Katrina.

John Glass, Ph.D., is a Professor in the JCVI Synthetic Biology Group. Glass is part of the Venter Institute team that recently announced the creation of a synthetic bacterial cell. In reaching this milestone the Venter Institute scientists developed the fundamental techniques of the new field of synthetic genomics including genome transplantation and genome assembly. His expertise is in molecular biology, microbial pathogenesis, RNA virology, and microbial genomics. At the JCVI he led the mycoplasma minimal genome, genome transplantation projects, and projects studying other mycoplasma and ureaplasma species. He has also participated in environmental genomics and viral metagenomics work. Glass and his Venter Institute colleagues are now using these and new synthetic genomics approaches to create cells and organelles with redesigned genomes to make microbes that can produce biofuels, pharmaceuticals, and industrially valuable molecules. Additionally, Glass is leading a new Venter Institute effort that uses synthetic genomics methods to improve the speed of production and efficacy of influenza virus vaccines. Glass is an adjunct faculty member of the University of Maryland at College Park Cellular and Molecular Biology Program. Prior to joining the JCVI Dr. Glass spent 5 years in the Infectious Diseases Research Division of Eli Lilly. There he directed a hepatitis C virology group and a microbial genomics group (1998–2003). Glass earned his undergraduate (biology) and graduate degrees from the University of North Carolina at Chapel Hill. His Ph.D. work was on RNA virus genetics in the laboratory of Gail Wertz. He was on the faculty and did postdoctoral fellowships in the Microbiology Department of the University of Alabama at Birmingham in polio virology with Casey Morrow and mycoplasma pathogenesis with Gail Cassell (1990–1998). On sabbatical leave in Ellson Chen's lab at Applied Biosystems Inc. (1995–1997) he sequenced the genome of *Ureaplasma parvum* and began his study of mycoplasma genomics.

Basil Golding, M.D., was born in Johannesburg, South Africa. He attended WITS Medical School and received his M.B. B.Ch. in 1968. He specialized and is board certified in Internal Medicine and Rheumatology. He performed research as a fellow at NIH and FDA. In 1995 he was appointed Chief, branch of Plasma Derivatives, and in 2004 Director, Division of Hematology, OBRR/CBER. He has published over 50 papers in peer-reviewed immunology journals and is currently involved in Toll-like receptor research.

Hana Golding, Ph.D., is the Chief of the Laboratory of Retrovirus Research at the Division of Viral Products, Center for Biologics Evaluation and Research (CBER), FDA. Dr. Golding received her bachelor's degree at the University of Jerusalem, Israel, and her Ph.D. degree from Oregon Health Sciences University. She also received postdoctoral training at the Experimental Immunology Branch of the National Cancer Institute. Dr. Golding has authored more than 100 research papers and book chapters on immunology, virology, and infectious diseases topics. Main areas of expertise are vaccines against viral pathogens including HIV, smallpox, and influenza, adjuvants mode of action, impact on immune responses, and biomarkers of in vivo toxicity.

Jesse L. Goodman, M.D., M.P.H., became Chief Scientist and Deputy Commissioner for Science and Public Health of FDA in 2009. He has broad responsibility for and engagement in leadership and coordination of the agency's crosscutting scientific and public health efforts. From 2003 to 2009, he was Director of FDA's Center for Biologics Evaluation and Research (CBER), which oversees medical and public health activities critical to U.S. and global preparedness concerning the development, evaluation, safety, quality, and availability of biologics. A graduate of Harvard, he received his M.D. from the Albert Einstein College of Medicine and did residency and fellowship training at the Hospital of the University of Pennsylvania and at UCLA (where he was also Chief Medical Resident). Prior to joining FDA, he was Professor of Medicine and Chief of Infectious Diseases at the University of Minnesota, where he directed the multihospital infectious diseases research, training, and clinical programs, and where his NIH-funded laboratory first isolated and characterized *Anaplasma phagocytophilum,* the infectious agent causing a new tick-borne disease, human granulocytic ehrlichiosis. He has authored numerous scientific papers and edited the book *Tick Borne Diseases of Humans* published by ASM Press in 2005. Dr. Goodman has been elected to the American Society for Clinical Investigation and to the Institute of Medicine of the National Academy of Sciences, where he is a longstanding member of the Forum on Emerging Threats. He is an active clinician and teacher who is board certified in Internal Medicine, Oncology and Infectious Diseases and is Staff Physician and Infectious Diseases Consultant at both the National Naval and Walter Reed Army Medical Centers, and is Adjunct Professor of Medicine at the University of Minnesota.

GIGI KWIK GRONVALL, PH.D., is a Senior Associate at the Center for Biosecurity of UPMC and an Assistant Professor of Medicine at the University of Pittsburgh. She is a term member of the Council on Foreign Relations, serves on the American Association for the Advancement of Science (AAAS) Committee on Scientific Freedom and Responsibility, and has been selected to participate in the European Union Visitors Programme for 2011. Dr. Gronvall's work addresses the role of scientists in biodefense—how they can diminish the threat of biological weapons and how they can contribute to an effective technical response against a biological weapon or a natural epidemic. Dr. Gronvall served as the Science Advisor of the Commission on the Prevention of Weapons of Mass Destruction Proliferation and Terrorism from April 2009 until the commission ended in February 2010. She is an Associate Editor of the quarterly journal *Biosecurity and Bioterrorism: Biodefense Strategy, Practice, and Science*. Dr. Gronvall received a B.S. in biology from Indiana University, Bloomington, and a Ph.D. from Johns Hopkins University.

RICHARD J. HATCHETT, M.D., is Chief Medical Officer and Deputy Director for Strategic Sciences and Management at the Biomedical Advanced Research and Development Authority (BARDA) within the U.S. Department of Health and Human Services. His primary responsibilities include oversight of programs relating to strategic science and innovation, strategic affairs and reporting, the development of science and preparedness policy, human resources, communications, and organizational marketing. Previously, he served as Director for Medical Preparedness Policy on the White House National Security Staff where he worked on a wide array of issues related to medical countermeasures development, the 2009 H1N1 pandemic, and pandemic preparedness more broadly. In 2005–2006, he served as Director for Biodefense Policy on the White House Homeland Security Council and was a principal author of the National Strategy for Pandemic Influenza Implementation Plan. In this capacity, he helped set policy and devise strategies to mitigate the consequences of a pandemic and promote pandemic preparedness. From 2005 to 2011, he served as Associate Director for Radiation Countermeasures Research and Emergency Preparedness at the National Institute of Allergy and Infectious Diseases. Dr. Hatchett completed his undergraduate and medical educations at Vanderbilt University, an internship and residency in Internal

Medicine at New York Hospital–Cornell Medical Center, and a fellowship in Medical Oncology at the Duke University Medical Center.

Judith Hewitt, Ph.D., is Chief of the Research Resources Section in the Office of Biodefense Research Affairs, DMID/NIAID/NIH. Her group is responsible for several resources within the Division of Microbiology and Infectious Diseases, namely contract programs providing animal models of infectious diseases, microbiology and infectious diseases biological resource repository, and in vitro assessments for antimicrobial activity. These programs all build upon successful prior efforts dedicated to biodefense, which are now being broadened to provide critical infrastructure capabilities for all pathogens of interest to DMID. Dr. Hewitt earned her Ph.D. from Johns Hopkins University and completed postdoctoral fellowships at NICHD and the Cleveland Clinic Foundation before taking positions at the University of Maryland and NIAID, where she ran transgenic and knockout mouse facilities focused on developing models for immunological studies.

Sally Hojvat, Ph.D., M.Sc., currently holds the position of Director of the Division of Microbiology Devices, in the Office of In Vitro Diagnostic Device (IVD) Evaluation and Safety, at FDA. Prior to joining FDA in 2003, Dr. Hojvat's experience included 18 years in the U.S. IVD industry where she held positions in the areas of IVD development and support, quality control, scientific affairs, and clinical research. Dr. Hojvat received a B.Sc. (Hons) from the University of Wales, UK, a M.Sc. in Microbiology from the University of Alberta, Canada, and a Ph.D. in Biochemistry from Loyola University Medical School, Chicago. She also completed postdoctoral training fellowships in Clinical Chemistry from Loyola Medical School and Pharmacology from the University of Chicago.

Robert House, Ph.D., is President of DynPort Vaccine Company LLC (DVC), a biotechnology firm in Frederick, Maryland, that manages product development programs for U.S. government agencies and provides consulting and technical and program management services to the biotechnology and pharmaceutical industries. The DVC portfolio includes innovative solutions for public health threats, particularly vaccines and therapeutics to protect against emerging infectious diseases, biological and chemical warfare threat agents, and seasonal and pandemic influenza. Prior to joining DVC, Dr. House worked at Covance Laboratories in Madison, Wisconsin, and IIT Research Institute in Chicago, Illinois, where he managed programs in immunotoxicology and safety assessment of pharmaceutical and biotechnology products. He has more than 20 years of experience in biomedical research and development, specializing in the assessment of inadvertent and therapeutic immunomodulation. Dr. House earned his Master of Science in Public Health and Ph.D. degrees in Medical Parasitology from the University of North Carolina at Chapel Hill School of Public Health. He is the author or co-author of more than 100 journal articles, book chapters and books covering immunology, toxicology, infectious disease and biodefense, and has participated in numerous working groups and expert panels in the field of immunotoxicology and biodefense. He is a frequent speaker at scientific conferences and university courses, and is a board member of the Tech Council of Maryland.

Jean Hu-Primmer, M.S., is a Senior Advisor for Regulatory Policy in the Office of Counterterrorism and Emerging Threats, Office of the Chief Scientist, U.S. Food and Drug Administration (FDA). She serves as Acting Director of the FDA Medical Countermeasures Initiative (MCMi) Regulatory Science, where she is leading the strategic development and coordinating the implementation of FDA's MCMi Regulatory Science program. Ms. Hu-Primmer began civil service career as a as a researcher-reviewer in the Division of Viral Products, Office of Vaccine Research and Review, Center for Biologics Evaluation and Research (CBER), FDA. During the early years of her tenure with the federal government, Ms. Hu-Primmer worked as a research scientist both at FDA CBER and at the Centers for Disease Control and Prevention in the area of influenza vaccine development and the pathogenesis of influenza viruses, including influenza A/H5N1 respectively. In the latter capacity, she worked to better understand host-virus interactions and the genetic/antigenic determinants of virulence specific to highly pathogenic avian influenza (HPAI) viruses. Ms. Hu-Primmer furthered her federal career as a regulatory affairs scientist for the National Institutes of Allergy and Infectious Diseases, Division of Microbiology and Infectious Diseases, and her responsibility included management of the entire influenza vaccine regulatory portfolio for the Division-sponsored clinical trials. Her regulatory submissions for the clinical development of the H5N1 vaccine provided the basis for the licensure of the sanofi pasteur H5N1 vaccine in the U.S. pandemic vaccine stockpile. In her current capacity, Ms. Hu-Primmer continues to serve as an Agency-wide subject matter expert related to influenza and influenza countermeasures as the chair of the FDA Pandemic Flu Task Force. She is the Agency focal point coordinating all influenza activities across the medical product Centers, and is the Agency representative to various influenza working groups constituted within the federal government. Likewise, as acting Director for the FDA MCMi Regulatory Science, she speaks on behalf of the Agency regarding the progress of various regulatory science projects and programs. Ms. Hu-Primmer holds a B.S. in microbiology from Cornell University and a M.S. in cellular and molecular biology from the University of West Florida, completing a postgraduate defense regarding the enhancement of the degradation of aromatic chlorinated hydrocarbons using a cloned bacterial hemoglobin gene in *Burkholderia cepacia*.

Patrick Iversen, Ph.D., was born in Carlsbad, New Mexico. He attended Westminster College with a major in Biological Sciences, completed his Ph.D. in Biochemical Pharmacology at the University of Utah School of Medicine, and completed postdoctoral training at the Eppley Institute for Cancer Research in Omaha. His experience in drug development extends back nearly 20 years with the initial seven years as a professor of pharmacology at the University of Nebraska Medical Center (UNMC). Thereafter, he worked on programs from 1990 to 1997 involving RNA therapeutics for cancer with leukemia as a primary area of investigation. He left UNMC in 1997 and led programs in cardiovascular disease, drug metabolism, muscular dystrophy, and antiviral drug development at AVI BioPharma. He developed broad experience in nearly every project from drug discovery, intellectual property protection, hypothesis development and study design, publication of scientific findings, preclinical toxicology design and evaluation, and preparation of IND and clinical protocols. He has served on multiple NIH study sections, reviewed for 37 different journals, and consulted more than a dozen different biotechnology companies.

Pravin R. Jadhav, Ph.D., FCP, is Team Leader and Expert Regulatory Scientist in the Division of Pharmacometrics of the Office of Clinical Pharmacology at the U.S. food and Drug Administration (FDA). He has worked on aspects of exposure–response to aid in important regulatory decisions such as drug–drug interactions, dose adjustment in special populations, evidence of effectiveness, benefit/risk, and labeling issues. He has several publications in peer-reviewed journals and presentations at international conferences. He has received several awards and honors at FDA, including the FDA Outstanding Service Award in 2008. Dr. Jadhav received his BPharm and MPharm from India, and a Ph.D. in Pharmaceutical Sciences from the Medical College of Virginia Commonwealth University in May 2006. He is a Fellow of the American College of Clinical Pharmacology.

J. Kevin Judice, Ph.D., is President and CEO, Achaogen. In addition to setting corporate strategy and building the leadership team, as President and CEO, Dr. Judice has also served as the company's Chief Scientific Officer. In this role, he has been responsible for the overall scientific strategy, research priorities, and portfolio management. Previously, Dr. Judice held positions of increasing responsibility at Genentech and Theravance. As one of the first 10 employees at Theravance, he played a key role in defining the company's scientific strategies and led the team that discovered telavancin, a new antibiotic for the treatment of multidrug-resistant "superbugs." Dr. Judice graduated magna cum laude with a B.S. in Chemistry from Texas A&M University. Following that, he received his Ph.D. in Organic Chemistry from UCLA and was an NIH postdoctoral fellow at UC Berkeley. In 2008, Dr. Judice was selected as a Henry Crown Fellow by the Aspen Institute. He is an author and inventor on over two dozen scientific papers and patents.

Michael Katze, Ph.D., has over 30 years of experience as a virologist and is a leader in applying genomic and proteomic technologies to the study of virus–host interactions and the innate immune response. He is an author of over 220 papers and reviews, of which over 50 are related to the use of functional genomic approaches to study virus–host interactions. He heads a laboratory of over 35 individuals and has considerable administrative credentials and experience managing large research endeavors. Dr. Katze is Program Director of a NIDA P30 Center focused on using genomic and proteomic technologies to study hepatitis C virus (HCV) and HCV-associated liver disease. This Center includes 23 key personnel and 23 other significant contributors from over a dozen academic institutions and three corporations. In addition, Dr. Katze serves as Associate Director at the Washington National Primate Research Center and is Head of the Center's Division of Functional Genomics and Infectious Disease, which is focused on developing genomic and proteomic resources for use in nonhuman primate research. In October 2008, Dr. Katze was awarded a NIAID contract to use systems biology approaches to develop computational models of the host response to respiratory virus infection.

George Korch, Ph.D., is a Senior Science Advisor for the Principal Deputy Assistant at the Office of Assistant Secretary for Preparedness and Response for the U.S. Department of Health and Human Services and is a Visiting Professor in the Department of Molecular Microbiology and Immunology at the Johns Hopkins Bloomberg School of Public Health. Dr. Korch retired from the U.S. Army Medical Department in 2008, where he had served in a number of leadership roles, including the Commander of the U.S. Army Medical Research Institute of Infectious Diseases and the Director of the Department of Defense Medical Chemical and Biological Defense Research Program. He also served as one of the first Directors of the National Biodefense Analysis and Countermeasure Center, Department of Homeland Security. His area of expertise is in viral and rickettsial zoonotic diseases and in medical countermeasure development (vaccines, therapies, and diagnostics) for biodefense needs. He serves or has served on such committees as the Institute of Medicine's Forum on Microbial Threats, the State of Maryland's Life Sciences Advisory Board, and with the Standards Development Committee for the American Type Cell Culture. Dr. Korch attended Boston University where he earned a B.S. in biology in 1974, followed by postgraduate study in mammalian ecology at the University of Kansas from 1975 to 1978. He earned his Ph.D. in immunology and infectious diseases from the Johns Hopkins School of Hygiene and Public Health in 1985, followed by postdoctoral experience at Johns Hopkins from 1985 to 1986.

Michael Kurilla, M.D., Ph.D., is the director of the Office of Biodefense Research Affairs and associate director for Biodefense Product Development for the National Institute of Allergy and Infectious Diseases (NIAID). His primary role is to provide overall institute coordination for product development of medical countermeasures against bioterror threats. At the University of Virginia, he was an assistant professor of pathology as well as co-director of the Laboratory of Molecular Diagnostics and associate director for clinical microbiology. Dr. Kurilla moved to the private sector working in anti-infective drug development at Dupont Pharmaceuticals, Bristol-Myers Squibb, and Wyeth. He subsequently joined NIAID as a medical officer. In 2005, he was named to his current positions within NIAID. He received his undergraduate degree in chemistry from the California Institute of Technology. He earned his M.D.-Ph.D. from Duke University. Dr. Kurilla took his postgraduate medical training in pathology at the Brigham & Women's Hospital in Boston, Massachusetts, and a postdoctoral fellowship with Dr. Elliott Kieff at Harvard Medical School as a Life Sciences Research Foundation fellow, followed by a Markey Scholar Award.

Elizabeth Leffel, Ph.D., M.P.H., is the Director of Non-Clinical Sciences for PharmAthene, Inc. In this role, she oversees research to ensure licensure of biodefense medical countermeasures under the FDA Animal Rule. Prior to joining PharmAthene, Dr. Leffel was the Senior Aerobiologist at the National Biodefense and Analysis Countermeasures Center where her primary responsibility was the establishment of an aerosol capability in Biosafety Level 3/4 laboratories. As the Chief of the Aerosol Research Center and Principal Investigator at the U.S. Army Medical Research Institute of Infectious Diseases (USAMRIID), her research focused on developing inhalation animal models for a number of biothreat agents. Dr. Leffel has served on the design committee for a BSL-3/4 lab, participated in progressing a BSL-3/4 lab from construction to preoperational level and served on biosafety committees for different institutions. She has published numerous manuscripts in peer-reviewed journals and written two book chapters. In 2007, she was earned a Department of Defense Science Technology Engineering and Math Award for her directorship of educational programs. Dr. Leffel received a Ph.D. in Toxicology and Pharmacology from the Medical College of Virginia, a M.P.H. from The George Washington University, and a B.S. degree in Biochemistry from Virginia Tech.

C. Rick Lyons, M.D., Ph.D., is currently the Director for Colorado State Infectious Disease Center. For the last 16 years Dr. Lyons was in the Department of Internal Medicine at the University of New Mexico Health Science Center in Albuquerque, New Mexico. He was Professor of Medicine and Director of the Center for Infectious Disease and Immunity. His research during that time focused on the application of molecular biology techniques to animal models of infection in order to move therapeutics for infectious diseases from basic science into human use. Over the last 10 years his work has focused on developing animal models for multiple biothreats in a variety of species. Dr. Lyons received his B.S. in Biochemistry from Washington State University in 1976; his Ph.D. in Immunology and Microbiology at University of Texas Southwestern in Dallas, Texas, in 1981, and subsequently his M.D. from UT Southwestern in 1987. He performed his internship, residency and fellowship in Hematology/Oncology at the Brigham and Women's Hospital in Boston, Massachusetts. He has received funding from NIH, DARPA, and the Department of Defense related to studying different aspects of host pathogenesis in various animal models of infection.

Colonel Christian Macedonia, M.D., is a military physician currently serving as the Medical Sciences Advisor to the Chairman of the Joint Chiefs of Staff. Dr. Macedonia works primarily in support of the chairman's efforts to improve the quality and availability of the best medical services America has to offer our Forces. He works broadly with interagency partners through his association with the National Security Council and with the nation's top research universities in this endeavor. His particular areas of published research include medical simulation, advanced medical imaging, fetal developmental biology, expedition medicine, and telemedicine. Dr. Macedonia is a Fellow of the Explorers Club recognized for his work on altitude medicine on Mt. Everest with NASA as well as the marine archeological expedition to the *RMS Titanic* in 2000. Dr. Macedonia holds a bachelors degree in chemistry from Bucknell University, a doctor of medicine degree from the Uniformed Services University, and he completed his residency in obstetrics and gynecology at Madigan Army Medical Center. He completed fellowship training at Georgetown University in maternal-fetal medicine followed by a research fellowship in bioinformatics at the National Institutes of Health. Colonel Macedonia is the recipient of many military awards and commendations including the Combat Action Badge and the Bronze Star Medal for his service in Iraq's Anbar Province as the Deputy Commander and Chief of the Medical Staff of the 115th Combat Support Hospital. His other honors include membership in the Alpha Omega Alpha Honor Medical Society, The Heroes of TRICARE Award, The Bockman Award, The NIH Award for Clinical Trainees, The Skelton Award, The "A" Proficiency Designator, and the Order of Military Medical Merit.

Alan J. Magill, M.D., FACP, FIDSA, is a Program Manager at the Defense Advanced Research Projects Agency (DARPA) where he initiates innovative and disruptive technology research and development programs. His current portfolio includes a $100M influenza vaccine program. He recently retired from 27 years active duty service in the U.S. Army. He was formerly the Director of the Division of Experimental Therapeutics at the Walter Reed Army Institute of Research (WRAIR) in Washington, D.C. Dr. Magill is ABIM board certified in internal medicine and infectious diseases. He has spent the last 20 years developing new generations of vaccines, diagnostics, and drugs directed against malaria and leishmaniasis. He has lived and worked in South America, Africa, and Southeast Asia. Dr. Magill previously served as the Head of Parasitology at the Naval Medical Research Center Detachment (NMRCD) in Lima, the Head of Clinical Research for the Malaria Vaccine Development Unit of the U.S. National Institutes of Health, and the Science Director for WRAIR. He has dual academic appointments as Associate Professor of Medicine and Associate Professor of Preventive Medicine and Biometrics at the Uniformed Services University of the Health Sciences (USUHS) in Bethesda, Maryland. He is a faculty member for the Gorgas Course in Clinical Tropical Medicine in Lima, Peru, continues to be a sought-after speaker on travel and tropical medicine-related topics, and a participant in numerous national and international advisory committees and workshops. He is an active member of the American Society of Tropical Medicine and Hygiene, serving as a Councilor in 2003–2006, the current Courses Director, and is the immediate Past President of their Clinical Group. Dr. Magill is the current (2009–2011) President of the International Society of Travel Medicine (ISTM) where he has been a member since 1992, serving as the Associate Chair of the Scientific Program Committee at CISTM9 (2005) in Lisbon and at CISTM10, (2007) in Vancouver. He is the Lead Editor of the 9th edition of *Hunter's Tropical Medicine*, the premier clinical textbook of clinical tropical medicine. He is also a Medical Editor of the CDC *Health Information for International Travel* (the yellow book) for 2010 and 2012. He has authored more than 65 peer-reviewed publications, 125 abstracts, and 13 book chapters.

Murray Malin, M.D., M.B.A., is the Acting Director of the Medical Countermeasure initiative at the Center of the Devices and Radiological Health (CDRH), U.S. Food and Drug Administration (FDA). A graduate of the University of Michigan, Dr. Malin received his M.D. from the Medical College of Ohio and did his residency and fellowship training in Anesthesiology and Critical Care at the Georgetown University Hospital in Washington, DC. While practicing anesthesiology, Dr. Malin earned his M.B.A. from the George Washington University in 2002, and began working as a Medical Officer in the Office of Compliance at CDRH in 2007, in which he was responsible for assessing risk and classification of recalls associated with defective medical devices.

Kenneth Mandl, M.D., M.P.H., has innovated and published extensively in the areas of personally controlled health records, disease outbreak detection, bio- and pharmacosurveillance, and national health information infrastructure. Recognized for his teaching and research, he has received the Barger Award for Excellence in Mentoring at Harvard Medical School and the Presidential Early Career Award for Scientists and Engineers, the highest honor bestowed by the United States government to outstanding scientists and engineers. Mandl co-directs a CDC Center of Excellence in Public Health Informatics. He is a leader of the SMArtPlatforms project—part of a major federal initiative seeking to create an "app store" for health. Mandl is a member of the Advisory Committee to the Director of the CDC and of Lister Hill Center Board of Scientific Counselors at the National Library of Medicine. He is an attending physician in pediatric emergency medicine, a faculty member in the Harvard Medical School Center for Biomedical Informatics, and affiliated faculty at the Harvard-MIT Division of Health Sciences and Technology.

Lisa Mathis, M.D., is a Pediatrician who attended the University of California, Davis, as an undergraduate with a B.S. in Physiology, and then attended the Uniformed Services University of Health Sciences (USUHS), F. Edward Hebert School of Medicine, and completed her Pediatric Residency at UC Davis. She has worked at FDA for 12 years and currently serves as the Director for the Pediatric and Maternal Health Staff in the Office of New Drugs. She also practices general pediatrics at the National Naval Medical Center in Bethesda, Maryland, and serves as an adjunct professor of medicine at USUHS. She is also the Deputy Team Leader on a rapid deployment team.

Susan McCune, M.D., is the Deputy Director in the Office of Translational Sciences (OTS) in the Center for Drug Evaluation and Research (CDER) at the Food and Drug Administration (FDA). She joined the

Agency in 2003 in CDER's Office of Counter-Terrorism and Pediatric Drug Development, Division of Pediatric Drug Development. From 2005 through 2009, Dr. McCune held the positions of Associate Director and team leader in the Office of Counter-Terrorism and Emergency Coordination. She joined OTS in February 2010. OTS comprises the Office of Biostatistics, the Office of Clinical Pharmacology, and the Immediate Office, which provides oversight to CDER research involving human subjects, CDER regulatory science research, and the CDER Computational Science Center. OTS is responsible for providing coordination for Critical Path initiatives across CDER in partnership with individual CDER offices. Dr. McCune received her medical degree from George Washington University following her undergraduate degree at Harvard University. She completed her internship, residency, chief residency, and neonatal fellowship at Children's National Medical Center in Washington, D.C. She is board certified in Pediatrics and Neonatal/Perinatal Medicine. For 15 years, while practicing academic pediatric and neonatal medicine at Johns Hopkins and Children's National Medical Center, Dr. McCune continued her molecular biology research on adrenergic receptor ontogeny and expression in models of newborn brain injury in the Lab of Developmental Neurobiology, NICHD, NIH. In addition, she has a Masters in Education Technology Leadership from George Washington University, and certificates in Public Health from Georgetown and Regulatory Science from USC.

Donna L. Mendrick, Ph.D., is the Director of the Division of Systems Biology at the National Center for Toxicology Research (NCTR), a research arm of FDA. Her division incorporates genomics, proteomics, metabolomics, bioinformatics, and *in silico* modeling approaches to answer the needs of FDA in terms of drug and food safety and improving the understanding of human disease. Her FDA committee assignments include the Senior Science Council, Critical Path Steering Committee, Tox21, and the Interagency Coordination Committee on the Validation of Alternative Methods (ICCVAM). Dr. Mendrick is a member of the Society of Toxicology's Disease Prevention Task Force. She was an Assistant Professor of Pathology at Harvard Medical School and Brigham and Women's Hospital. She joined Human Genome Sciences and, as a Group Leader in Pharmacology, oversaw multiple project teams, toxicity studies, pharmacology studies, etc. Prior to joining FDA, she was a Scientific Fellow and Vice President of Pharmacogenomics at Gene Logic where she oversaw pharmacogenomics and spearheaded its toxicogenomics effort. For the latter, she formed a pharmaceutical consortium to help guide the development of the program. Dr. Mendrick has over 25 years of experience in the fields (in alphabetical order) of immunology, pathology, pharmacogenomics, pharmacology, toxicology, and toxicogenomics employing small molecule drugs, recombinant therapeutic proteins, and monoclonal antibodies. Dr. Mendrick has published on the use of pharmacogenomics, metabolomics, and proteomics to identify biomarkers. She currently is a committee member of the Predictive Toxicology Discussion Group at the New York Academy of Sciences and is past President of the National Capital Area Chapter of the Society of Toxicology. Dr. Mendrick was on the Editorial Board of the *Journal of Histochemistry and Cytochemistry* for 8 years, a member of the NIH SBIR Immunology Study Section for 8 years, and a member of the Board of Directors of the National Kidney Foundation of Massachusetts for 4 years.

Nancy Messonnier, M.D., obtained her M.D. degree from the University of Chicago and completed a residency in internal medicine at the University of Pennsylvania. Dr. Messonnier joined CDC in 1995 as an Epidemic Intelligence Service Officer in the Meningitis and Special Pathogens Branch. Dr. Messonier's research has focused on bacterial meningitis and other vaccine preventable diseases (including *Neisseria meningitidis, Haemophilus influenzae,* and *Bordatella pertussis*) and bacterial zoonoses in the United States and internationally, evaluation and development of vaccines, and surveillance for infectious diseases. She is currently responsible for the CDC Anthrax Vaccine Research Program (AVRP) and co-author for the recently published ACIP guidelines for use of anthrax vaccine. In 2001, she played a leadership role the field investigation of the first identified bioterrorism-related case of *Bacillus anthracis* in Florida. She made critical contributions in the field in Washington, DC, as well as in evaluation of the overall epidemiology of the outbreak. She led the evaluation of antimicrobial postexposure prophylaxis for *B. anthracis* among 10,000 individuals exposed to *B. anthracis,* as Primary Investigator, developed a new protocol for postexposure prophylaxis with antibiotics and anthrax vaccine for persons exposed to *B. anthracis* spores in the event of a new attack.

Linda A. Miller, Ph.D., is the Director of Clinical Microbiology in Infectious Diseases Medicines Discovery and Development at GlaxoSmithKline (GSK) Pharmaceuticals. Dr. Miller joined GSK in 1994. She obtained her Ph.D. in 1987 from the University of Pennsylvania, and also has a Master's Degree from the Medical College of Pennsylvania. Dr. Miller directs the Clinical Microbiology group at GSK that includes Anti-bacterial and Anti-viral drug development and has responsibilities across the Infectious Diseases pipeline from discovery to development and throughout the life cycle of the drugs. Her focus at GSK includes clinical microbiology, antimicrobial resistance, resistance modeling, surveillance, and science policy.

Her previous experience includes her role as Director of Clinical Microbiology and Clinical Immunology at Holy Redeemer Hospital and Medical Center in Pennsylvania and her position as the Clinical Immunologist for The Bryn Mawr Hospital, Bryn Mawr, Pennsylvania. Dr. Miller is the current chair of Division A, Antimicrobial Agents and Chemotherapy, for the American Society for Microbiology (ASM) and also is a member of the ASM's International Membership Committee. She was President of the Eastern Pennsylvania Branch of the ASM from 1992 to 1994, and has recently co-chaired two symposia for the Eastern PA Branch of the ASM. Her research and specialty areas include global antimicrobial surveillance systems, antimicrobial susceptibility testing, in vitro methodologies, susceptibility testing breakpoints, bacterial identification methodologies, immunofluorescence science policy, and drug life cycle management.

Robert C. Nelson, Ph.D., FISPE, was a career Public Health Service officer with 21 years at the U.S. Food and Drug Administration. He served as a new drug reviewer, drug abuse scheduling expert, epidemiology team leader, Director of the Center for Drug Evaluation & Research (CDER) Staff College, and as Associate Director, Office of Epidemiology & Biostatistics, before his retirement in 1998. He was responsible for the reengineering of the CDER postmarketing program and designed, managed and implemented the Adverse Reaction Reporting Systems (AERS). Dr. Nelson also led the comprehensive regulatory rewrite of all safety regulations ("The Tome") in the United States, and ensured ICH compatibility.

Robert M. "Skip" Nelson, M.D., Ph.D., is currently the Senior Pediatric Ethicist/Lead Medical Officer in the Office of Pediatric Therapeutics, Office of the Commissioner at the U.S. Food and Drug Administration. After receiving his M.D. degree from Yale University, Dr. Nelson trained in pediatrics (Massachusetts General Hospital), neonatology, and pediatric critical care (University of California, San Francisco). He has a Master of Divinity degree from Yale Divinity School and a Ph.D. in The Study of Religion from Harvard University. Dr. Nelson is a former Chair of the FDA Pediatric Advisory Committee and the Pediatric Ethics Subcommittee. He was a member of the Subcommittee on Research Involving Children of the Secretary's Advisory Committee on Human Research Protections, and the Human Studies Review Board of the Environmental Protection Agency. Dr. Nelson was a member of the Committee on Clinical Research Involving Children of the Institute of Medicine, and former Chair of the Committee on Bioethics of the American Academy of Pediatrics. Dr. Nelson is the editor of the *American Journal of Bioethics* (AJOB)–Primary Research, which publishes empirical research in bioethics. Immediately prior to joining FDA, he was Professor of Anesthesiology, Critical Care and Pediatrics, at the Children's Hospital of Philadelphia and University of Pennsylvania School of Medicine. Dr. Nelson's academic research explored various aspects of child assent and parental permission, including risk perception and voluntary choice, and was funded by the Greenwall Foundation, the National Institutes of Health, and the National Science Foundation.

Ed Nuzum, D.V.M., Ph.D., During the time Dr. Nuzum served in the U.S. Army from 1971 to 1998 he was a member of the Army Transportation Corps and the Army Veterinary Corps and held positions as diverse as Platoon Leader, Aircraft Maintenance Officer, and Pentagon Medical Staff Officer. While in the Army he served as a helicopter pilot, he received the D.V.M. degree from Kansas State University in 1982 and his Ph.D. degree from the University of Kansas Medical Center in 1990, and held research and management positions at the U.S. Army Medical Research Institute of Infectious Diseases and the U.S. Army Walter Reed Army Institute of Research. From 1998 to 2002 Dr. Nuzum was a Senior Scientist and Product Development Team Leader with DynPort Vaccine Company where he worked on vaccinia immune globulin and vaccines for smallpox and Venezuelan equine encephalitis. From 2002 to present Dr. Nuzum has been with the Office of Biodefense Research Affairs (OBRA), a part of the NIAID Division of Microbiology and Infectious Diseases. In 2004 he was made Chief of the Biodefense Vaccines and other Biological Products Development Section (BVBPDS), which is responsible for planning, implementing, and evaluating advanced biologics product development efforts involving extramural research contracts and interagency agreements. A major focus at NIAID has been the development of anthrax vaccines and associated animal models. Dr. Nuzum has over 30 years experience in veterinary clinical, medical research, and research management experience.

Gerald Parker, D.V.M., Ph.D., M.S., is the principal deputy assistant secretary to the assistant secretary for preparedness and response. Since March 2003, he has been detailed to the Department of Homeland Security. During his career, he has held a variety of positions, including assistant deputy for research and development and research director for the Medical Chemical and Biological Defense Research Program at the U.S. Army Medical Research and Materiel Command. In this role, he led joint service and interagency programs responsible for developing research investment strategies and sustaining unique capabilities to develop a broad range of medical countermeasures. He is a former commander and deputy commander of the U.S. Army Medical Research Institute of Infectious Diseases, the lead DoD medical research laboratory for medical biological defense. In these positions, he directed the technology-based research and development of vaccines, diagnostics, and drugs, along with the development of medical defense strategies and the training of health care providers against biological warfare agents and highly infectious organisms requiring special containment. Dr. Parker graduated from Texas A&M University with a B.S. in veterinary medicine and a degree of doctor of veterinary medicine. He holds a doctorate in physiology from Baylor College of Medicine in Houston, Texas, and an M.S. degree in resourcing the national strategy from the Industrial College of the Armed Forces.

Vikram S. Patel, Ph.D., joined FDA in 2010 as a Deputy Director in the Division of Drug Safety Research in CDER. At FDA he is responsible for guiding safety-related preclinical research, including research in the area of toxicology, computational sciences, pharmacokinetics, drug metabolism, and transporters. Prior to joining FDA Dr. Patel was a Senior Director of Discovery Pharmacokinetics at Wyeth. He is currently serving as a member of the National Acadamies Committee for Animal Models for Assessing Countermeasures to Bioterrorism Agents. Dr. Patel has extensive experience in drug discovery and development. He has expertise in the areas of pharmacokinetic/pharmacodynamic modeling and simulations, physiological modeling (including biomarker modeling and simulations), drug metabolism, in vitro/in vivo correlations, and in-drug formulation and delivery. He developed and established a GLP preclinical PK section at Procter and Gamble Pharmaceuticals and developed Macrobid, a sustained release product currently marketed worldwide. Dr. Patel received his Ph.D. from the University of Houston in 1984.

Carl Peck, M.D., obtained a B.A. in mathematics and chemistry from the University of Kansas in 1963 and the M.D. in 1968. Following training in internal medicine, he undertook a research fellowship in clinical pharmacology at the University of California San Francisco (1972–1974). From 1974 to 1980, Dr. Peck was employed at the Letterman Army Institute of Research, San Francisco, California, as Chief of the Army Blood Preservation Research Program. In 1980, Dr. Peck became Director of the Division of Clinical Pharmacology and, Professor, Departments of Medicine and Pharmacology, Uniformed Services University, Bethesda, Maryland. Dr. Peck joined FDA as Director, Center for Drug Evaluation and Research, in October 1987. He was promoted to Assistant Surgeon General in the Public Health Service in October 1990. Retiring from FDA in late 1993, Dr. Peck was appointed "Boerhaave" Professor of Clinical Drug Research at Leiden University in The Netherlands. In 1994 Professor Peck joined the faculty of the Georgetown University Medical Center, as the founding Director of the Center for Drug Development Science. In 1999, Dr. Peck received the FDA Distinguished Alumnus Award. Sweden's University of Uppsala conferred an honorary doctorate degree (Doctor Honoris Causa) to Dr. Peck in January 2002 in recognition of "outstanding contributions to the science of drug development." Dr. Peck founded NDA Partners LLC in 2003, and in 2004, CDDS moved to UCSF, located in the UC-Washington Center. Throughout his career, he has mentored more than 40 postdoctoral fellows and graduate students and co-founded the American (2007) and Chinese (2009) Courses in Drug Development and Regulatory Science (ACDRS, CCDRS). Dr. Peck's research interests center on optimizing informativeness, efficiency, speed, and economy of drug development and regulation using advanced concepts and techniques of clinical pharmacology, trial designs, and pharmacostatistical modeling and simulation to generate causal evidence of effectiveness and safety. He is an author of more than 150 original research papers, chapters, and books.

Mary K. Pendergast, J.D., LL.M., is President of Pendergast Consulting, which provides legal and regulatory advice to biopharmaceutical companies, patient groups, professional and advocacy organizations, governments, and academic and financial institutions. Ms. Pendergast professional focus is on strategic and tactical issues that relate to drug and device policy and development. Prior to her current position, she was Executive Vice President Government Affairs of Elan Corporation. She has held positions as a corporate officer, devising and implementing regulatory strategies for product development and compliance; was liaison to BIO and PhRMA; and testified for BIO before Congress on PDUFA reauthorization. She was Deputy Commissioner and Senior Advisor to the Commissioner at the Food and Drug Administration involved in FDA's efforts to regulate emerging areas, such as biotechnology, cellular and tissue-based therapies, genetic testing, xenotransplantation, and acute-care research, and served as FDA's "crisis manager," handling sensitive and precedent-setting situations. She also held the position of Associate Chief Counsel for Enforcement at the FDA Office of the General Counsel in which she supervised a wide variety of enforcement and defensive litigation involving FDA programs and products under FDA's jurisdiction; and was an attorney at the Office of the General Counsel, Department of Health and Human Services and as special assistant to the Department's General Counsel. Ms. Pendergast is on the boards of directors of ARCA biopharma, Inc., AesRX, and the Arch Foundation.

John H. Rex, M.D., received his M.D. degree from Baylor College of Medicine (1982), trained in Internal Medicine at Stanford University

Hospital (1984–1987), and trained in infectious diseases at the National Institute of Allergy and Infectious Diseases (1987–1992). John served on the faculty of the University of Texas Medical School at Houston from 1992 to 2002 during which time his work focused on laboratory studies of novel antifungal agents, clinical trials of novel antifungal agents, and hospital epidemiology. John joined AstraZeneca in 2003, and he currently serves as Vice President, Clinical Infection. In addition to his AZ role, he is the industry representative on the FDA Anti-Infective Drug Advisory Committee, is Chair of the Area Committee on Microbiology for the Clinical Laboratory Standards Institute (CLSI, formerly NCCLS), is a Highlights Advisor for Nature Reviews Microbiology, serves on several editorial boards, was formerly an editor for *Antimicrobial Agents and Chemotherapy*, and is an Emeritus Editor for www.doctorfungus.org, a nonprofit website devoted to dissemination of information about medical mycology.

Eric Rose, M.D., is an academic physician and entrepreneur with interests in drug discovery, biodefense, clinical evaluative research, and health policy. Since 2007 he has been the Executive Vice President for Life Sciences at MacAndrews & Forbes and CEO of Siga Technologies, Inc., a developer of antiviral drugs directed at potential agents of bioterror. He was appointed in 2007 to the National Biodefense Scientific Board, which advises the HHS Secretary on biodefense, influenza, and emerging diseases. In 2008, he assumed the chairmanship of the Department of Health Policy at the Mount Sinai School of Medicine. From 1994 through 2007, he served as Surgeon in Chief at New York-Presbyterian Hospital/Columbia and Chairman of the Department of Surgery at the Columbia University College of Physicians and Surgeons, where he held a distinguished professorship. An accomplished heart surgeon, researcher, and entrepreneur, Dr. Rose grew one of the nation's premier departments of surgery while managing, investigating, and developing complex medical technologies ranging from heart transplantation and novel approaches to Alzheimer's disease to bioterrorism. He has authored or co-authored more than 300 scientific publications and has received more than $25 million in NIH support for his research. Dr. Rose pioneered heart transplantation in children, performing the first successful pediatric heart transplant in 1984, and has investigated many alternatives to heart transplantation, including cross-species transplantation and man-made heart pumps. Siga has received more than $100 million in federal research support since he joined the company, developing antiviral drugs for smallpox, dengue, and Lassa fever. He received both his undergraduate and medical degrees from Columbia University.

Stephen Ruberg, Ph.D., received his B.A. in mathematics from Thomas More College, an M.S. in statistics from Miami University (Ohio), and his Ph.D. in biostatistics from the University of Cincinnati. Steve has spent 30 years in the pharmaceutical industry and just completed his 11 year anniversary at Lilly. During his career, he has served as a statistician for all phases of drug development from discovery through postmarketing. He has worked on drug development programs across numerous therapeutic areas. In 1994, Steve was elected as a Fellow of the American Statistical Association, and he has published widely in statistics and biological/medical journals. He has held significant leadership roles across the industry: Deputy Chair and co-author of ICH-E9–Statistical Principles for Clinical Trials; chaired PhRMA Biostatistics and Data Management Committee; helped found CDISC (the Clinical Data Interchange Standards Consortium) and served as its first Chairman of the Board; chaired the Board of Governors for the Ohio State University Mathematical Biosciences Institute; and was appointed to the Board or Directors of the National eHealth Collaborative. In 2009, he was named Scientific Leader for the Eli Lilly Advanced Analytics Hub.

Harvey Rubin, M.D., Ph.D., received his Ph.D. in Molecular Biology from the University of Pennsylvania in 1974 and his M.D. from Columbia University in 1976. He was a House Officer in Medicine at The Peter Bent Brigham Hospital in Boston and did his fellowship in infectious diseases at Harvard and the Brigham. He is currently Professor of Medicine at the University of Pennsylvania with secondary appointments in the Departments of Microbiology, Biochemistry, and Computer and Information Sciences. His research on the basic biology of tuberculosis and other bacteria and the mathematical modeling of complex biological systems has been funded by the NIH, NSF, DARPA, the Global Alliance for TB Drug Discovery, and the DoD. He has published over 90 papers in peer-reviewed journals as well as numerous scientific reviews and book chapters. Dr. Rubin served on a number of national and international scientific review panels including those for the NIH, NSF, NASA Intelligent Systems Program, DARPA, and the Medical Research Council, South Africa. He was a member of the United States National Science Advisory Board for Biosecurity (NSABB) and the Department of Defense/National Academy of Sciences Biological Cooperative Threat Reduction Program. He is the Chair, Scientific Advisory Board, Incentives for Global Health. Dr. Rubin is the Director of Penn's Institute for Strategic Threat Analysis and Response (ISTAR). ISTAR is dedicated to identifying, analyzing, and solving policy, scientific, and technical issues that contribute to regional, national, and international security.

Estelle Russek-Cohen, Ph.D., is the acting division director in the Division of Biostatistics in the Center for Biologics Evaluation and Research (CBER) in the Office of Biostatistics and Epidemiology. She came to CBER in February 2010 as Deputy Division Director. Before that she was a team leader in the Diagnostic Devices Branch of the Division of Biostatistics, in FDA's Office of Surveillance and Biometrics at the Center for Devices and Radiological Health. Dr. Russek-Cohen received a Ph.D. in Biostatistics from the University of Washington, Seattle. Dr. Russek-Cohen was a professor in the University of Maryland's Biometrics Program for 26 years and Director of the Biometrics Program for her last five years at College Park when she retired in 2004 and came to FDA. At UM, she regularly collaborated with scientists and epidemiologists on infectious disease research. She also spent a year of sabbatical leave and several summers at the Biometric Research Branch of the National Cancer Institute working on statistical issues in clinical trials. Her current interests in statistics include the assessment of safety of CBER-regulated products and statistical issues in personalized medicine. She is a fellow of the American Statistical Association.

Alan Shaw, Ph.D., joined VaxInnate from the Merck Vaccine Research Division where he was Executive Director of MVD's Public Policy, Public Health, and Medical Affairs Department. Prior to this, he was the Executive Director of Virus & Cell Biology at Merck Research Laboratories, responsible for all the aspects of live virus vaccine research, as well as technical aspects of development and production. His responsibilities covered research and early development of recombinant protein-based vaccines. Dr. Shaw was instrumental in the development of a combination measles-mumps-rubella-varicella vaccine (ProQuad), a live oral rotavirus vaccine, (RotaTeq), human papillomavirus vaccine (Gardasil), and zoster vaccine (Zostavax), as well as numerous early-stage experimental vaccines. He has over 15 years of experience in the development, testing manufacturing, and implementation of vaccines in the United States, Europe, and in international programs. Prior to joining Merck, Dr. Shaw worked on vaccines for hepatitis B and *Plasmodium falciparum* as well as cytokines, cell trafficking, and natural inhibitors of inteleukin-1 at Biogen, SA, in Geneva, Switzerland.

David Siegel M.D., FAAP, received his bachelor's degree in biological sciences from SUNY at Stony Brook, attended medical school at New York Medical College, and completed his pediatric residency at the Montifiore Hospital of the Albert Einstein School of Medicine. He has had a multifaceted career spanning 40 years as a clinician, educator, and administrator in academic and community-based settings. As a clinician he became one of the first EMTs in the country and worked as a pediatrician in the outpatient, emergency, and inpatient settings. As an educator/administrator he became an ambulatory attending at the Children's National Medical Center immediately following completion of his residency program; he then had a major role in the development and management of a new pediatric residency program at INOVA Fairfax Hospital for Children, and was the inpatient director at Connecticut Children's Hospital. Following the events of September 11, 2001, while working in the emergency department at Children's National Medical Center, he initiated and managed the development of a pediatric disaster preparedness education and training program (CBRNE focus), conducted through a grant from HRSA. Dr. Siegel joined the National Institute of Child Health and Human Development in October 2007. A major role of his functioning as a medical officer in the Obstetric and Pediatric Pharmaceutical Branch at NICHD has been the identification of WMD-related medical countermeasures gaps as well as the facilitation of the development of WMD-related medical countermeasures for pregnant women and children. He is the obstetric/pediatric representative from NIH on various Biomedical Advanced Research and Development working groups as well the NIH representative to the Federal Education and Training Interagency Group. As a child advocate he is a federal liaison to both the National Commission on Children and Disasters and the American Academy of Pediatrics' Disaster Preparedness Advisory Council.

Lt. Col. Daniel J. Wattendorf, M.D., USAF, joined DARPA as a Program Manager in the Defense Sciences Office in 2010. His interests focus on applying methodological advances in genomics and biotechnology to optimize health and prevent disease—specifically to achieve simple solutions that improve health care at the point of care, anywhere. He holds a B.S. in microbiology from Cornell University and a medical degree with distinction from George Washington University. He completed a residency in family medicine at the National Capital Consortium; a residency in clinical genetics at the National Human Genome Research Institute (NHGRI) at NIH; a fellowship in clinical cytogenetics at Georgetown University; and a fellowship in health policy from the Office of the Director, NHGRI, NIH. Lt. Col. Wattendorf previously served as Director, Air Force Medical Genetics Center, and program manager for an Advanced Concept Technology Demonstration integrating advanced diagnostics and informatics with surveillance systems to rapidly detect natural and hostile pathogens in the Office of the Air Force Surgeon General. In addition to his DARPA programs, he is a geneticist at the National Naval Medical Center and the Cancer Genetics Branch, National Cancer Institute, NIH.

JEFFREY WETHERINGTON, PH.D., attended Rutgers University from which he earned a Ph.D. in statistics. He is currently at GlaxoSmithKline where he is directing a group that provides statistical analysis and modeling, and clinical trials simulation expertise to early-phase drug discovery and development activities for new medicines targeted at infectious diseases.

CAROLYN WILSON, PH.D., received her Ph.D. in Genetics from George Washington University while working in the laboratory of Dr. Robert Gallo for her dissertation research. For her postdoctoral fellowship, she worked in the laboratory of Dr. Maribeth Eiden identifying viral and cellular factors influencing viral entry. She joined the Division of Cellular and Gene Therapies at the FDA Center for Biologics Evaluation and Research in 1993. As a researcher-reviewer in DCGT, she reviewed INDs and developed policy and guidance documents in two novel product areas: gene therapy and xenotransplantation. More recently, Dr. Wilson has served as the Associate Director for Research at CBER. As ADR, Dr. Wilson ensures that CBER's research is relevant, of high quality, and provides CBER with the appropriate scientific expertise, tools, and data to support regulatory decision making and policy development. Dr. Wilson still maintains her own laboratory program studying retroviruses that are either used as vectors for gene therapy clinical trials or are of concern in the xenotransplantation setting.